MAKE YOUR
AIRPLANE
LAST FOREVER

BY NICHOLAS E. SILITCH

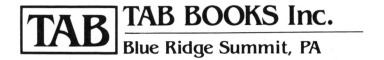

TAB BOOKS Inc.

Blue Ridge Summit, PA

FIRST EDITION

THIRD PRINTING

Printed in the United States of America

Library of Congress Cataloging in Publication Data

Silitch, Nicholas E.
    Make your airplane last forever.

    Includes index.
    1. Airplanes—Maintenance and repair. I. Title.
TL671.9.S53   629.134′ 6   82-5690
ISBN 0-8306-2328-0 (pbk.)   AACR2

TAB BOOKS Inc. offers software for sale. For information and a catalog, please contact TAB Software Department, Blue Ridge Summit, PA 17294-0850.

Questions regarding the content of this book should be addressed to:

    Reader Inquiry Branch
    TAB BOOKS Inc.
    Blue Ridge Summit, PA 17294-0214

Contents

Acknowledgments

The aviation industry's help in writing this book reflects its interest in the promotion of aviation and aviation safety, as well as its interest in the reduction of the high cost of flying.

Particular attention must be called to the help of Jack R. Knott, Field Service Engineer for Edo-Aire, and for his unusual approach to saving money in the operation of avionics, as reflected in Chapter 4. In addition, more special thanks are due for his photographs and the personal tutoring on the subject.

Ken Johnson, of Avco Lycoming, Williamsport Division, furnished much of the information (and all of the photographs) used in Chapters 2 and 3. Without Ken's help, these chapters would have been considerably less informative.

Last, but not least, thanks to my wife, Mary Frances, who found time from her aviation writing at *AOPA Pilot* magazine to help me over the rough spots.

Introduction

The methods and techniques in this book require some effort on the part of the pilot—at best, self-control and discipline; at worst, some grease under the fingernails and time spent doing things we might rather not do. If the effort is made, however, and the advice is followed with a reasonable amount of diligence, you will spend less on maintenance and repairs of the engine, avionics and other components. You will enjoy a longer life from them and the rest of the airplane, while having an airplane that looks new and fresh for as long as you care to fly it.

Most of the data in this book was provided by the manufacturers of engines, avionics, airframes and accessories. In spite of what some may think, the manufacturers are all interested in seeing you fly longer, safer and cheaper with their products. In a case where manufacturers disagreed, I have sided "with the evidence." (An example is the long-argued matter of whether it is permissible to mix brands of oil. Lycoming says no, except when absolutely necessary; Continental thinks it's just fine. Although I have never had any trouble with mixing brands of oil, I gave the nod to Lycoming because of the consistently longer TBOs of their engines. The proof of the pudding is, after all, in the eating.)

John Wayne never had to worry about excessive engine cooling as he peeled off to attack the enemy, nor can I ever recall seeing him cleaning the chewing gum out of the carpets, but then he didn't have to pay the bills, either. Keeping the airplane clean and shiny won't make it fly much better, or make it any safer—only make it worth more to you, and the purchaser when that time comes. The information on the engine and avionics will not only save you money, but the increased reliability and your increased knowledge of the components will provide you with an extra edge of safety when the going gets rough, making you a better pilot in the bargain.

Chapter 1

Understanding the Engine

Regrettably, a basic understanding of the processes at work in the engine is necessary to obtain maximum utilization of it. I say *regrettably* because there are those of us who have no trouble understanding the workings of the internal combustion engine, and those of us who have long ago given up the attempt to understand chapters of books dedicated to this subject.

If you are one of the latter, stay with this chapter. Complete comprehension of the material presented is not necessary: only a vague idea of the operation of the engine is needed to understand what happens to an engine when or before unpleasant things occur. The information in this chapter emphasizes the areas where things are likely to go wrong when the engine is abused, as well as where and why "normal" wear occurs. *Normal* wear is defined as wear that is unavoidable by design, but sufficiently under control to give a reasonable life to the engine.

It is not the purpose of this chapter to convert the mechanically illiterate. It is not really necessary for the owner/pilot to understand exactly how the engine works, only what makes it *stop* working.

Airplane engines are, for purposes of this discussion, all four-stroke in design. Two-stroke engines are being used in some ultralight aircraft; they are the sort of engine for which you have to mix oil with the fuel. Although they may look externally similar to their four-stroke equivalents, they are quite different inside. They

1

Fig. 1-1. The way we were: a nine-cylinder radial engine.

do not have the dependability or longevity necessary for conventional aircraft applications. In addition, due to their internal design, their fuel consumption is very high in relation to the horsepower achieved.

Reciprocating aircraft engines come (or more accurately came) in three types, radial, in-line, and opposed. The latter type is the survivor. Radial engines are seldom seen in new airplanes; when they are, the airplane may be new, but the engine usually is not. They are seen mostly in cropdusters, antiques, and classics. Radial engines, due to their design, present a greater frontal area and therefore create more drag than an opposed engine. In addition, they are oil consumers and tend to be less efficient users of fuel than other types (Fig. 1-1).

The in-line engine had a brief moment of glory in the 1930s, but there were always problems with air-cooling the rear cylinders, as well as with the height of the engines. These design headaches did not exist with the opposed engine. When our current crop of airplanes was designed, the only viable choice was the air-cooled, opposed, four-stroke engine (Fig. 1-2).

BASIC THEORY

The internal combustion engine is a strange thing. Push the button, and the thing erupts; fuel goes in and smoke, noise, heat, and power come out. How does it work?

The Intake Stroke

During the intake stroke, the piston is pulled downwards by the rotation of the crankshaft, pulling the air and fuel mixture through the open intake valve. (How the intake valve happens to be open, and how the fuel got into the air being sucked into the cylinder, are covered later.) Obviously, if the piston is to pull the mixture into the cylinder, it must be a pretty good fit in the cylinder. If it isn't, it will suck air by the piston, rather than through the intake valve. To this end, the piston is fitted with piston rings (Fig. 1-3), whose job it is to prevent escape of gases by this route.

The Compression Stroke

As the crankshaft continues its rotation, the piston reverses direction at bottom dead center (BDC) and starts upwards. Shortly after this moment, the intake valve closes. The mixture is compressed by the piston in its upward travel and prevented from passing by the valves by their accurate fitting against the valve

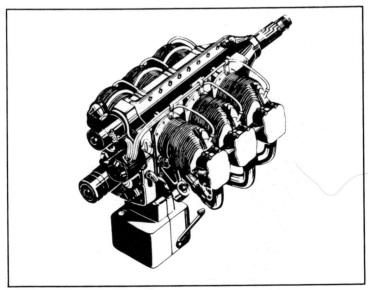

Fig. 1-2. The way we are: a six-cylinder horizontally opposed engine.

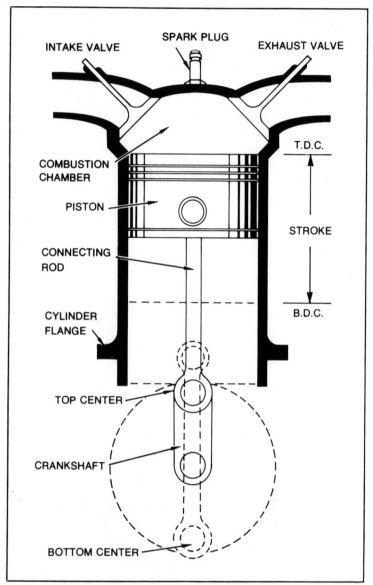

INTAKE VALVE

SPARK PLUG

EXHAUST VALVE

COMBUSTION
CHAMBER

PISTON

CONNECTING
ROD

CYLINDER
FLANGE

TOP CENTER

CRANKSHAFT

BOTTOM CENTER

T.D.C.

STROKE

B.D.C.

Fig. 1-3. Components and terminology of engine operation.

seats, and contained at the bottom by the piston rings. As the piston approaches the top of its travel (TDC), the spark plug is ignited. It, in turn, ignites the fuel, which (when all is functioning properly) burns smoothly away from the spark plug electrode (Fig. 1-4).

4

The Power Stroke

As the crankshaft continues its rotation, the piston again reverses its direction as it passes top dead center (TDC) and starts down. The ignition of the fuel, which started at or near the end of the compression stroke, continues rapidly, the flame moving away from the electrode and consuming all the mixture in the cylinder. The rapid expansion of the gases caused by their combustion pushes the piston downwards, as the gases are prevented from leaving the cylinder by the piston rings and valves. The power stroke is the only one of the four strokes that makes the crankshaft, and thus the engine, turn. During all the other engine strokes, the piston is being moved by the crankshaft.

Shortly before the piston reaches bottom dead center (BDC), the exhaust valve opens, and the exhaust gases start on their way out of the engine.

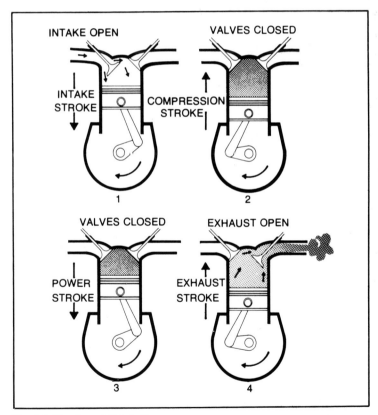

Fig. 1-4. The four-stroke cycle.

The Exhaust Stroke

As the piston passes bottom dead center (BDC) it is pushed upwards again by the crankshaft. As it moves upward, it continues pushing the exhaust gases out the open exhaust valve. Shortly before the piston reaches top dead center (TDC), the intake valve opens, allowing the mixture to start into the cylinder, completing the cycle. If you still are a little unsure of this process, go back to "The Intake Stroke" and read it through again, referring to the diagram, until you have at least a general idea of the process.

You will notice that in the description of the four-stroke cycle, the valves opened and closed mysteriously at the right time, the mixture was always there when it was supposed to be, and the spark plug happened to fire just when it was needed. Obviously, the timing of the valves and the ignition, and the amount and quality of the mixture, are very important.

VALVES

If you will remember in the basic description of the four-stroke cycle, the valves, oddly opened and closed a little bit after you might have thought they should. Why does the intake valve open while the piston is still traveling upwards, still pushing the exhaust gases out of the cylinder? Wouldn't this push the exhaust gases into the intake manifold? The answer to this question has to do with the same fluid theory that makes your airplane fly (Fig. 1-5).

As the exhaust gases are pushed out of the cylinder into the exhaust stacks or manifold, they increase in velocity due to the restriction presented. This velocity tends to suck, or scavenge, the remaining exhaust gases out of the cylinder, as well as suck the mixture into the cylinder. Allowing the exhaust valve to close a little bit after TDC and the intake valve to open before TDC takes advantage of this scavenging action.

A camshaft operates the valves through a tappet, pushrod, and rocker arm (to change the direction). The camshaft is connected by a gear to the crankshaft, ensuring the accuracy of the timing (Fig. 1-6).

In time, or with improper lubrication, the lobes on the camshaft (as well as other parts of the valve train) can wear. Wear of the camshaft is not desirable, and can only be repaired by replacement, while small amounts of wear on the rest of the valve train are normal, and may be compensated for by adjustment. Some engines have self-adjusting tappets which make adjustment unnecessary. Any wear of these components affects the very subtle timing of the

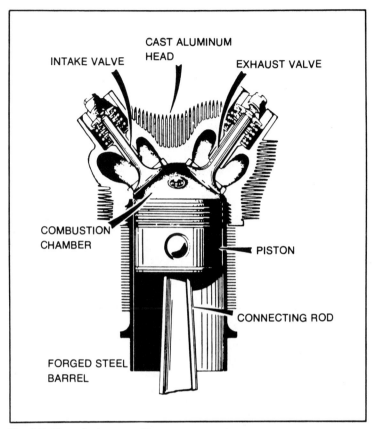

Fig. 1-5. The cylinder assembly.

valves, causing power loss, inefficiency, and possible valve damage (Fig. 1-7).

If the valves in your engine can be manually adjusted, it is important to do this by the book to avoid problems. If they are automatically adjusted, the system must be functioning correctly.

The valves are the front line of the engine. Understanding the four-stroke principle, it is easy to see why the seal they make against their seats must be perfect. Any leakage here will cause a loss of compression in the engine, and thus a loss of power. In addition, valves are very susceptible to heat damage, and an improper seat will cause them to overheat, because their primary cooling comes from their contact with their seats, which transfer heat from the valve to the head. Additional cooling is supplied by the valve guides.

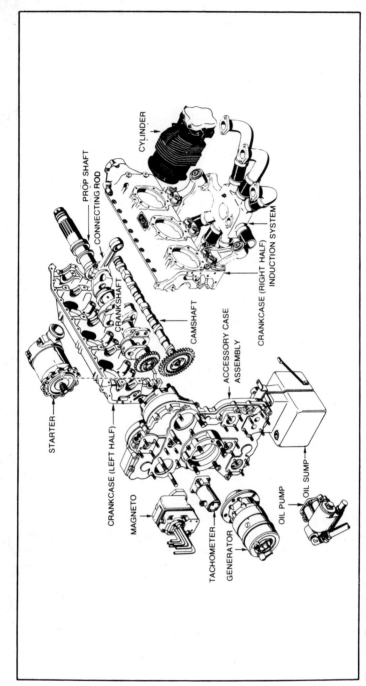

Fig. 1-6. Exploded engine view.

CYLINDER

PROP SHAFT

CONNECTING ROD

CRANKSHAFT

CAMSHAFT

INDUCTION SYSTEM

CRANKCASE (RIGHT HALF)

ACCESSORY CASE ASSEMBLY

STARTER

CRANKCASE (LEFT HALF)

MAGNETO

TACHOMETER

GENERATOR

OIL PUMP

OIL SUMP

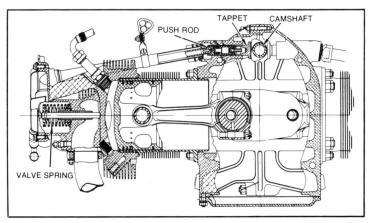

Fig. 1-7. Reciprocating parts of the opposed engine.

Since there is little lubrication in this environment, wear in the valve guides is considered normal. As it gets out of hand, it causes a deterioration of the heat transfer, in addition to allowing exhaust gases to blowby the valve stem and engine oil to be sucked into the combustion chamber on the intake side.

Heat damage affects valves in many ways, all of them ending in the same unpleasant way: failure. Therefore, it is vitally important to avoid this sort of damage. Once the valve guides are worn beyond tolerable limits, they should be replaced and the valves reseated (a top overhaul) to ensure their continued performance. There are several ways to determine when this is necessary. Occasionally a valve will burn, causing a poor seal. The usual cause of this is a poor seat, which causes the valve to overheat.

A word to the wise—valves are intolerant of excessively quick cooling. A sudden cooling of the valve from its normal, almost white-hot running temperature usually results in a slight warpage. This causes a deterioration in the seat and rapid wear of the guides.

The intake valves are cooled by the mixture flowing past them when they are open. Additionally, they are protected from the most intense heat produced by the engine by being closed against their seats at that time. For these reasons, they are usually pretty hardy, and most potential trouble occurs in the exhaust valves, which are open while the blowtorch heat of the exhaust gases passes by them. Exhaust valves and seats are made of special heat-resistant materials, and are filled with sodium in both head and stem of some, only in the stems of others) in all but the smallest engines, to aid in the transfer of heat to the head through the valve guide.

IGNITION

Aircraft ignition systems are duplicated partly for safety reasons and partly to ensure optimum combustion. That is to say, there are two spark plugs per cylinder, two sets of high-tension wiring, and two magnetos. On some new engines, the magnetos share a single drive gear and shaft. One purpose of this practice is to allow the engine to function if one side of the ignition expires for any reason.

The magnetos are gear-driven from the camshaft or crankshaft, and very carefully timed to deliver the spark at the best instant to ignite the mixture. You may have noticed a small anomaly in the description of the four-stroke cycle—the spark plug fires *before* the piston reaches top dead center on the intake stroke, before, in fact, the intake valve is fully closed. This is because the flame takes time to travel, and by the time that the expansion of gases has reached a substantial level, the piston has started its downwards power stroke, and both valves are tightly closed. This brief moment before top dead center, when the mixture has started burning while the piston is still moving up on the intake stroke, and the intake valve is still partially open, is called "spark advance", and is expressed in degrees before top dead center. This setting varies in different engines, and is critical to the operation of the engine.

Igniting the charge at this point permits maximum pressure to build up at a point slightly after the piston passes over top dead center. For ideal combustion, the point of ignition should vary with engine speed and with degree of compression, mixture strength, and other factors governing the rate of burning. However, certain factors, such as the limited range of operating rpm and the dangers of operating with incorrect spark settings, prohibit the use of variable spark control in most instances. Therefore, most aircraft ignition system units are timed to ignite the fuel/air charge at one fixed position (advanced).

On early models of the four-stroke-cycle engine, the intake valve opened at top center (beginning of the intake stroke). It closed at bottom center (end of intake stroke). The exhaust valve opened at bottom center (end of power stroke) and closed at top center (end of exhaust stroke). More efficient engine operation can be obtained by opening the intake valve several degrees before top center and closing it several degrees after bottom center. Opening the exhaust valve before bottom center, and closing it after top center, also improves engine performance. Since the intake valve opens before top-center exhaust stroke and the exhaust valve closes after top-center intake stroke, there is a period where both the intake and exhaust valves are open at the same time. This is known as valve lap or valve overlap. In valve timing, reference to piston or crankshaft position is always made in terms of before or after the top and bottom center points, *e.g.*, ATC, BTC, ABC, and BBC.

Opening the intake valve before the piston reaches top center starts the intake event while the piston is still moving up on the exhaust stroke. This aids in increasing the volume of charge admitted into the cylinder. The selected point at which the intake valve opens depends on the rpm at which the engine normally operates. At low rpm, this early timing results in poor efficiency since the incoming charge is not drawn into and the exhaust gases are not expelled out of the cylinder with sufficient speed to develop the necessary momentum. Also, at low rpm the cylinder is not well scavenged, and residual gases mix with the incoming fuel and are trapped during the compression stroke. Some of the incoming mixture is also lost through the open exhaust port. However, the advantages obtained at normal operating rpm more than make up for the poor efficiency at low rpm. Another advantage of this valve timing is the increased fuel vaporization and beneficial cooling of the piston and cylinder.

Delaying the closing of the intake valve takes advantage of the inertia of the rapidly moving fuel/air mixture entering the cylinder. This ramming effect increases the charge over that which would be taken in if the intake valve closed at bottom center (end of intake stroke). The intake valve is actually open during the latter part of the exhaust stroke, all of the intake stroke, and the first part of the compression stroke. Fuel/air mixture is taken in during all this time.

The early opening and late closing of the exhaust valve goes along with the intake valve timing to improve engine efficiency. The exhaust valve opens on the power stroke, several crankshaft degrees before the piston reaches bottom center. This early opening aids in obtaining better scavenging of the burned gases. It also results in improved cooling of the cylinders, because of the early escape of the hot gases. Actually, on aircraft engines, the major portion of the exhaust gases, and the unused heat, escapes before the piston reaches bottom center. The burned gases continue to escape as the piston passes bottom center, moves upward on the exhaust stroke, and starts the next intake stroke. The late closing of the exhaust valve still further improves scavenging by taking advantage of the inertia of the rapidly moving outgoing gases. The exhaust valve is actually open during the latter part of the power stroke, all of the exhaust stroke, and the first part of the intake stroke.

From this description of valve timing, it can be seen that the intake and exhaust valves are open at the same time on the latter part of the exhaust stroke and the first part of the intake stroke. During this valve overlap period, the last of the burned gases are escaping through the exhaust port while the fresh charge is entering through the intake port.

Many aircraft engines are supercharged. Supercharging increases the pressure of the air or fuel/air mixture before it enters the cylinder. In other words, the air or fuel/air mixture is forced into the cylinder rather than being drawn in. Supercharging increases engine efficiency and makes it possible for an engine to maintain its efficiency at high altitudes. This is true because the higher pressure packs more charge into the cylinder during the intake event. This increase in weight of charge results in corresponding increase in power. In addition, the higher pressure of the incoming gases more forcibly ejects the burned gases out through the exhaust port. This results in better scavenging of the cylinder.

Normal combustion occurs when the fuel/air mixture ignites in the cylinder and burns progressively at a fairly uniform rate across the combustion chamber. When ignition is properly timed, maximum pressure is built up just after the piston has passed top dead center at the end of the compression stroke.

The flame fronts start at each spark plug and burn in more or less wavelike forms. The velocity of the flame travel is influenced by the type of fuel, the ratio of the fuel/air mixture, and the pressure and temperature of the fuel mixture. With normal combustion, the flame travel is about 100 ft./sec. The temperature and pressure within the cylinder rises at a normal rate as the fuel/air mixture burns.

There is a limit, however, to the amount of compression and the degree of temperature rise that can be tolerated within an engine cylinder and still permit normal combustion. All fuels have critical limits of temperature and compression. Beyond this limit, they will ignite spontaneously and burn with explosive violence. This instantaneous and explosive burning of the fuel/air mixture or, more accurately, of the latter portion of the charge, is called detonation.

As previously mentioned, during normal combustion the flame fronts progress from the point of ignition across the cylinder. These flame fronts compress the gases ahead of them. At the same time, the gases are being compressed by the upward movement of the piston. If the total compression on the remaining unburned gases exceeds the critical point, detonation occurs. Detonation then, is the spontaneous combustion of the unburned charge ahead of the flame fronts after ignition of the charge.

The explosive burning during detonation results in an extremely rapid pressure rise. This rapid pressure rise and the high instantaneous temperature, combined with the high turbulence generated, cause a "scrubbing" action on the cylinder and the piston. This can burn a hole completely through the piston.

The critical point of detonation varies with the ratio of fuel to air in the mixture. Therefore, the detonation characteristic of the mixture can be controlled by varying the fuel/air ratio. At high power output, combustion pressures and temperatures are higher than they are at low or medium power. Therefore, at high power the fuel/air ratio is made richer than is needed for good combustion at medium or low power output. This is done because, in general, a rich mixture will not detonate as readily as a lean mixture.

Unless detonation is heavy, there is no cockpit evidence of its presence. Light to medium detonation does not cause noticeable roughness, temperature increase, or loss of power. As a result, it can be present during takeoff and high-power climb without being known to the crew.

In fact, the effects of detonation are often not discovered until after teardown of the engine. When the engine is overhauled, however, the presence of severe detonation during its operation is indicated by dished piston heads, collapsed valve heads, broken ring lands, or eroded portions of valves, pistons, or cylinder heads.

The basic protection from detonation is provided in the design of the engine carburetor setting, which automatically supplies the rich mixtures required for detonation suppression at high power; the rating limitations, which include the maximum operating temperatures; and selection of the correct grade of fuel. The design factors, cylinder cooling, magneto timing, mixture distribution, and carburetor setting are taken care of in the design and development of the engine and its method of installation in the aircraft.

The remaining responsibility for prevention of detonation rests squarely in the hands of the ground and flight crews. They are responsible for observance of rpm and manifold pressure limits. Proper use of supercharger and fuel mixture, and maintenance of suitable cylinder head and carburetor-air temperatures are entirely in their control.

Preignition, as the name implies, means that combustion takes place within the cylinder before the timed spark jumps across the spark plug terminals. This condition can often be traced to excessive carbon or other deposits which cause local hot spots. Detonation often leads to preignition. However, preignition may also be caused by high-power operation on excessively lean mixtures. Preignition is usually indicated in the cockpit by engine roughness, backfiring, and by a sudden increase in cylinder head temperature.

Any area within the combustion chamber which becomes incandescent will serve as an igniter in advance of normal timed ignition and cause combustion earlier than desired. Preignition may be caused by an area roughened and heated by detonation erosion. A cracked valve or piston, or a broken spark plug insulator may furnish a hot point which serves as a "glow plug."

The hot spot can be caused by deposits on the chamber surfaces resulting from the use of leaded fuels. Normal carbon deposits can also cause preignition. Specifically, preignition is a condition similar to early timing of the spark. The charge in the cylinder is ignited before the required time for normal engine firing. However, do not confuse preignition with the spark which occurs too early in the cycle. Preignition is caused by a hot spot in the combustion chamber, not by incorrect ignition timing. The hot spot may be due to either an overheated cylinder or a defect within the cylinder.

The most obvious method of correcting preignition is to reduce the cylinder temperature. The immediate step is to retard the throttle. This reduces the amount of fuel charge and the amount of heat generated. If a supercharger is in use and is in high ratio, it should be returned to low ratio to lower the charge temperature. Following this, the mixture should be enriched, if possible, to lower combustion temperature.

MIXTURE

The proper mixture arrives at the intake valve through the intake manifold. It consists of air and fuel. The fuel is mixed with the air at the carburetor or fuel-injector nozzle. In either case, the result is the same. Fuel is mixed with filtered air and metered in suitable quantities to the engine.

In all modern aircraft, the mixture may be leaned or richened by the operator. This is not the only area in the fuel system that is operator governable, but it is probably the most important. Excess leaning at low altitudes and under heavy loads (takeoff and climb) can damage the engine. Excessively lean engines run very hot and can damage pistons and valves. At maximum loads, the engine is running hot already, and cannot absorb the additional heat without damage. If the engine is running too hot in climb (or even cruise), particularly at low altitudes, the mixture should be richened and the throttle reduced.

If there are leaks in the manifold, throttle butterfly bushings, or other places in the intake system, they may lead to an excessive leaning, as well as allowing unfiltered air into the combustion chamber. Neither of these conditions can be tolerated by the engine for long.

Handling of the throttle has a great deal to do with the longevity of the engine, and is fully covered in a later chapter. Suffice it to say that *smoothness* is the key here—sharp changes of speed in the engine, or excessive power demands from a cold engine can cause major damage.

COMBUSTION CHAMBER

The combustion chamber is where all the action is. The parts are not as subject to damage as the valves, as cooling is readily accomplished when everything is adjusted properly. Lubrication of the cylinder walls and rings, is however, a compromise at best.

Rings and Walls

It is very important that the seal between the piston and cylinder wall is in perfect condition. Since it is impossible to make a piston that would fit perfectly into the bore of the cylinder under all operating conditions, the seal is not accomplished by the piston itself, but rather by rings mounted on the piston (Fig. 1-8). The

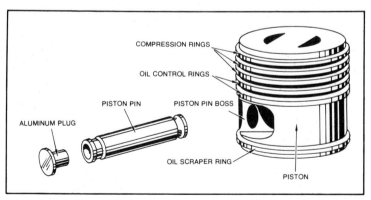

Fig. 1-8. The piston assembly.

piston in the illustration has five rings—the actual number may differ, but there are always at least three.

The ring in the upper groove is the compression ring. This ring holds the compression in the cylinder during the power cycle, as well as maintaining the suction in the chamber during the intake cycle. These are frequently slightly different in design from each other when there is more than one. This allows the designer to have an individual ring accomplish each secondary purpose of the compression ring, while doubling or tripling the primary compression function.

The cylinder wall is lubricated with oil, either by splash, spray, or pressure through the piston pin. This oil must be strictly controlled to avoid allowing excess amounts of it into the combustion chamber where it will foul plugs, create carbon deposits, and cause excessive oil consumption. This control is accomplished by the oil control ring or rings that scrape away the excess oil and leave a thin, uniform quantity on the cylinder wall. It should leave just enough oil on the cylinder wall to avoid metal-to-metal contact.

The bottom-most ring on the piston is the oil scraper ring. Its function is to carry oil up the cylinder wall on the upward stroke of the piston, so the control ring or rings have some oil to distribute on the way down.

It will be seen that each of the different types of rings has an important function in which exact fit is critically important. There is a gradual wear of the rings where they scrape the cylinder wall, as well as wear on the piston grooves from the constant up and down movement of the rings against them. In addition to the wear on the piston and grooves, the cylinder wall is also wearing. The piston rings maintain great pressure against the walls of the cylinder, and in doing so tend to wear that surface.

The piston rings are designed to take up this wear, which is considered normal. Abnormal wear in the cylinders comes from dirt entering the combustion chamber, either from the air in the mixture or the oil. As a matter of fact, most of the dirt in the oil enters the engine with the mixture and is cleaned away from the combustion chamber by the oil, which is polluted in the process. The dirt in the combustion chamber wears the rings, piston grooves, and cylinder walls abnormally fast. In small quantities, the abnormal wear caused by faulty filtration will take many hours to show up as reduced compression and increased oil consumption, so it is best to guard against it by scrupulous maintenance of the air filtration system. Spectroscopic examination of the oil will catch this problem before it can cause too much damage by showing the increased levels of silica (sand) in the oil as well as the increased levels of metal (from the worn surfaces).

Ingestion of large quantities of dirt or a foreign object can gouge the sides of the cylinders, causing the same sort of damage, but instantaneously. If a valve breaks, it can cause cylinder wall damage as it knocks around the combustion chamber, although it is just as likely to impale the piston. Piston rings can break through improper fitting and improper operating technique, which will score the cylinder wall, causing a compression failure and excessive oil use.

As the engine will be considered run-out when the piston rings can no longer fulfill their functions, close monitoring of the oil consumption is recommended—it can give you an accurate measure of how things are with the rings and cylinder walls.

Pistons and Connecting Rods

Pistons are manufactured of aluminum for both its light weight and its superior heat transfer capabilities. There is normally little wear in pistons; but if anything goes wrong in the combustion chamber, it is likely to mess up the piston, as it is the softest thing around as well as being the most subject to melting or cracking.

Because the piston is made of aluminum (which has a very high rate of expansion with the application of heat), and has to fit reasonably well into a cylinder bore which is made of steel (which has a very low coefficient of expansion), the piston's fit must be designed for operating temperatures. This means that when the pistons and cylinder walls are cold, the fit is looser than when the engine is at running temperature. If the engine is started and put under full load while still cold, the piston will get very hot, expanding, while the cylinder wall stays cold. This can lead to a seizing of the piston in the cylinder. Usually, this sort of seizing is not felt at the time it occurs, but it will score the piston as well as the cylinder. This is why it is important to warm the engine properly, particularly in cold weather. A scuffed cylinder or piston must be replaced.

Some engines are more prone than others, to piston groove wear which requires replacement of pistons as well as their rings at rebuild time. This wear is caused by the constant changes of direction of the piston and the consequent movement of the piston ring in the groove. This sort of wear is seldom the cause for rebuilding an engine, although oil control rings with excessive clearance can act as pumps, pumping excess quantities of oil into the combustion chamber. If this situation gets out of hand, the pistons and rings will have to be replaced.

BEARINGS

Plain bearings are usually used for the critical bearings in modern airplane engines. These are the crankshaft main bearings (mains), the crankshaft rod bearings (rods) and the camshaft bearings. I call them critical because they, along with the cylinder and ring condition, determine the useful life of the engine (Fig. 1-9).

In modern engines, these bearings are of the insert type. A steel shell is lined with an alloy bearing surface, usually an alloy of lead, tin, copper or antimony. It is not designed to run metal against metal; it must have a layer of oil between the relatively soft bearing surface and the crankshaft journal. If there is no oil here, the bearing will fail quickly. Therefore, oil delivery to these bearings is very important.

That is why these bearings are usually the first place the oil goes once it leaves the filter and pump (Fig. 1-10). The oil reaches the bearings through galleries drilled in the crankshaft. These give very good lubrication to the bearings, which is why they will run a long time without noticeable wear. Once wear exceeds normal, however, they will wear very quickly. The usual ways to judge this wear on an engine is by oil inspection and/or analysis and oil pressure monitoring. The oil inspection will disclose small pieces of bearing material in the oil as the bearing starts on this accelerated wear phase of its life, as will the spectroscopic oil analysis. The

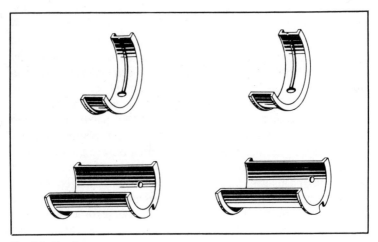

Fig. 1-9. Bearings.

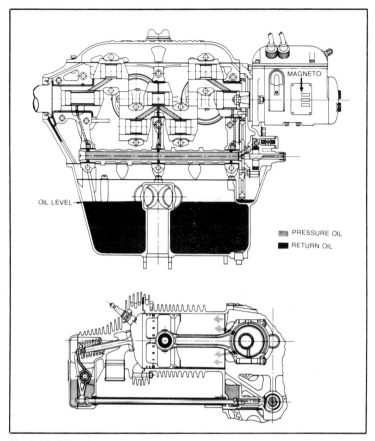

OIL LEVEL

MAGNETO

PRESSURE OIL
RETURN OIL

Fig. 1-10. Oil system.

spectroscopic analysis will, of course, catch the problem sooner, as it will disclose a minute rise in the bearing alloy quantity in the engine.

The use of oil pressure monitoring is somewhat more complex. The oil pressure of the engine is determined by the restriction caused by the allowable clearance between the bearing and the journal. As this clearance increases over the life of the engine, the oil pressure tends to drift downwards. This is considered normal. What is *not* normal is the sudden drop in oil pressure caused by excessive opening of this clearance. A sudden drop (but it can still be in the green) in oil pressure frequently signals an imminent bearing failure. Even a speeding up of the usual slow, downward trend in oil pressure will usually indicate that the end of the useful

19

life of the engine is near. Invariably, in a high-time engine, this sort of acceleration in oil pressure loss can be confirmed to be bearing wear by checking the oil filter or a spectroscopic analysis.

Dirt in the oil is the one thing that will significantly shorten the life of these bearings. As a rule, they are still in pretty good shape when the engine reaches its normal TBO; they are replaced prophylactically.

Chapter 2

Avoiding Premature Engine Demise

Airplanes have gotten considerably more practical than they used to be, even though they might have lost a great deal of their romantic appeal in the process (Fig. 2-1). In fact, the best news about the present pedestrian generation of aircraft is their sheer reliability and minimized maintenance. This is reflected in the high time between overhauls (TBO) of modern aviation powerplants, which are expected to run 1500 to 2000 hours, against the 300 hours or so hoped for from the engine of the '30s. Bear in mind that Lindbergh's historic solo flight from New York to Paris was most notable for the fact that his Wright Whirlwind engine functioned nonstop for 33 hours.

RELIABILITY

Where has the added reliability of the modern aviation engine come from? Credit cannot be given to any one area; the process has been gradual, a chipping away at all the little problems that added up to the big one. The results have been gratifying: a 600 percent increase in TBOs, and an even greater increase in reliability. The work has been done over the years in three major areas: metallurgy, lubricants, and filters. Without judging which has contributed most, we will discuss them in that order.

Metallurgy

The science of metallurgy goes back to the first primitive who

21

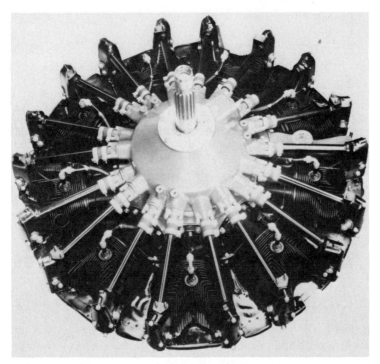

Fig. 2-1. Avco Lycoming R-680. (Courtesy Avco Lycoming)

discovered that mixing some tin with his molten copper gave him an alloy that was harder and more useful then either copper or tin by itself. This ushered in the Bronze Age, a period that has confused many of us in the early years of our educations and convinced us that history is a subject best left to others to study.

The Iron Age soon followed, and (particularly in the last century) techniques of hardening, softening and casting it developed rapidly. In the middle of the last century, steel was developed—a necessity for the progress of the industrial revolution, and, of more interest to us, the internal combustion engine (without which flying would have been impossible). From its earliest development, progress was constant to the point that today iron and steel can be manufactured for an infinite variety of purposes. In recent years we have learned how to surface treat steel and iron to combine desired characteristics, such as the strength necessary in the crankshaft and the hardness desirable in the journals to minimize wear.

Cylinder barrels are nitrided or chrome plated to harden the surface that the piston rings work against, while steel surfaces may be etched to retain oil or prevent corrosion. A better knowledge of the properties and strength of metals has allowed engineers to design parts to closer tolerances, shaving weight and resulting in lighter, stronger, and longer-wearing parts. Hand-poured and scraped babbitt bearings have been replaced with precision insert bearings, steel shells with exotic alloys to carry the loads. These are lighter, longer-wearing, precise, and easy to replace. Valve seats and valves are made of alloys and combinations of metals undreamed of fifty years ago, allowing leaner mixtures and more efficient operation, which in turn result in increased engine life, economy, and reliability.

Lubricating Oils

We'll limit our mention of engine lubricants to mineral oils. The synthetics will undoubtedly take over sometime in the future because it seems certain that they will be superior in every way to mineral oils. Meanwhile, we'll continue to fly with the old standbys.

There are two kinds of aircraft engine oil for recips, those with a single viscosity such as grade 40 or grade 65, and the multi-grade oils such as SAE 20W-40. The former is usually referred to as a "straight mineral oil," while the latter is often called a "compound oil." Actually, both kinds contain additives. The so-called "compound oil" or multi-grade oil will be an ashless dispersant (AD) oil, and it is definitely superior to the single-grade oil when it comes to protecting your engine.

In very warm weather, use of the multi-grade or AD oil will probably result in a slight drop in oil pressure along with a small increase in oil temperature. However, it has a greater film strength, is blended to operate at higher temperatures, and tests have established that there is no detectable difference in engine wear between engines lubricated with AD and straight mineral oils—except that since most engine wear occurs during start-up and the first few seconds of engine operation, the AD oil significantly reduces engine wear during this period because, cold, it flows through the system much quicker. For all practical purposes, the 20W-40 is a 20 weight oil when cold, and a 40 weight oil when hot. And it would be an appropriate lubricant for an engine in which the manufacturer had recommended a grade 80 aviation oil.

We can see that we'd better pause here and explain the grading system because it can be confusing. First, some aviation oils are graded the same way as automotive oils, and these will have SAE numbers. Others simply double the SAE viscosity. For example,

Grade 80 aviation oil is the same viscosity as SAE 40. Grade 40 is the same as SAE 20. But just to keep you from being complacent about it, Grade 65 is equal to SAE 30, not 32½. (You've heard the old saying that holds that there's a right way, a wrong way, and the Army way. So you won't be surprised to learn that the military adds 1000 to the civil aviation oil grades so that MIL 1065 is the same as Grade 65—and SAE 30.)

The average plane owner isn't going to learn very much about the additives that go into av-oils because the several brands are in competition regard their formulas as trade secrets, and therefore are prone to describe their products in terms of what they do for you rather than offer a detailed list of ingredients—which most of us wouldn't understand anyway.

But the important things are clear. The so-called "straight mineral" oils do not have as much film strength as the newer AD oils—which is why you are told to break-in a freshly overhauled engine on straight mineral, the theory being that an AD oil gives too much protection and therefore does not allow the rings to seat properly.

And while on the subject of break-in after overhaul, do not baby a freshly overhauled engine! That initial wear—most of which will take place during the first hour of operation—is very important because it rounds off the sharp grooves in the new piston rings and establishes the seal between rings and cylinder wall. This seating process continues, to an ever-decreasing degree, for perhaps the first 100 hours. By then the oil consumption should have stabilized and the degree of success (or luck) you've had in obtaining an optimum "working relationship" between the new rings and the newly ground cylinder walls will be apparent.

Not every overhaul job will turn out perfect; and most veteran av mechanics will agree, we think, that the reasons can seldom be pinned down. The best, most experienced mechanic can do it the same way every time, yet find that the results are not exactly the same every time.

Is it safe to switch to a multi-viscosity AD oil in an engine that has been operated for many hundreds of hours or more with straight mineral oil? There is no good reason why one can't switch to an AD oil from straight mineral. If there is reason to believe that the engine may be heavily sludged, it may be wise to check the oil screens and the filter after about ten hours to make sure the cleaning action of the AD oil hasn't taken into suspension contaminants that might possibly restrict the oil circulation. However, these cleaning agents are not harsh and the cleaning action is really a rather slow if steady process.

Oil Filters

There are a few pilot/owners around who believe that regular oil filter changes, along with oil replenishment due to normal consumption, is all that is needed to insure good engine protection. This practice seems to be based on the belief that oil doesn't wear out, it merely gets dirty. The US Government once issued a statement to that effect; and the military "rerefined" aircraft engine oil for years.

We now know that oil does wear out. Its molecules are not only damaged by extreme heat, but also tear apart, and therefore film strength deteriorates.

So, even if you changed oil filters every day you couldn't—or shouldn't—avoid your regular 50-hour oil change. Also, no filter so far developed can trap all the contaminants, the smaller particles are just too small.

Just remember, you don't get something for nothing. As compression ratios go up you increase the load on your engine oil. The extreme force with which your piston compression rings— particularly the top one—are pressed against the cylinder wall is proportionate to the engine's compression ratio, because the burning gases in the combustion chamber, driven into the piston's ring grooves behind the rings, expand the rings against the cylinder wall with great pressure (perhaps 1,000 lbs sq/in), while the oil film between ring and cylinder wall completes the seal and prevents destructive metal-to-metal contact. Both extreme heat and pressure are at work here to attack this critical oil film.

We therefore end up with some very simple and easy-to-follow guidelines when it comes to aircraft engine oils: Use a top brand—it doesn't matter which one—of a multi-viscosity (say, SAE 10W-30 or SAE 20W-40) and change it, along with the oil filter, every 50 hours of engine operation under normal conditions. The best engine oil you can buy is a great bargain compared to the cost of engine repair. Use a straight mineral oil for break-in after overhaul, and do not baby the engine during this period.

FIVE BASICS FOR MAXIMUM ENGINE LIFE

Joe Diblin of Avco Lycoming compiled this list of five basics for achieving maximum engine life, and I can't think of a better source of expertise. Lycoming, who made the engines for the legendary

Dusenberg and Cord automobiles, turned their knowledge to aviation in 1928 and have developed an enviable reputation in this field. Their reputation is for consistently maintaining longer TBOs than their competition; in fact, it might be said that if you want to save money owning an airplane, the best way to start might be buying one with a Lycoming engine (Figs. 2-2 through 2-6).

Air Filter

As a result of Lycoming's long experience in building and operating aircraft engines, they have concluded that the induction air filter is a very important element in the life of an aircraft engine. With the modern high-performance powerplant, the operator must keep dirt and abrasives out of the engine if he is to attain the expected life and trouble-free hours of operation. Lycoming observed an excellent example of the importance of filtering when a twin-engine pilot told them that his last oil analysis indicated a very high amount of silicon. They advised the pilot to operate on filtered air *only* for the next 50 hours, and not to use the ram (unfiltered) position. After 50 hours on filtered air only, another analysis showed that the iron content had lowered considerably, while the silicon content had dropped to normal.

Excessive wear and early failure of reciprocating engine parts is due, in many instances, to contaminents introduced through or around the air filter. If a worn or poorly fitted air filter allows as much as a tablespoon of abrasive material in the cylinders, it will

Fig. 2-2. Avco Lycoming 0-235. (Courtesy Avco Lycoming)

Fig. 2-3. Avco Lycoming 0-360. (Courtesy Avco Lycoming)

cause wear to the extent that an overhaul may be required. Evidence of dust or other material in the induction system beyond the air filter is indicative of inadequate filter care or a damaged filter or housing.

To prevent undesirable combustion chamber wear, follow the instructions outlined in your aircraft and engine manuals on filter maintenance procedures. They will stress such recommendations as inspecting the entire induction system to preclude the introduction of unfiltered air between the filter and the fuel distribution system. The manual will also recommend inspecting the carburetor heat door, or the alternate air door, to ensure that they are working properly. Any accumulation of dirt and dust in or near the entrance of the alternate air or carburetor heat doors will be drawn into the engine when the doors are opened. Aircraft parked in dusty areas should be carefully inspected in these areas before each use.

Oil

Clean engine oil is essential to long engine life, and the full flow oil filter is a great improvement over older methods of filtration. Generally, service experience has shown that the use of external oil filters can increase the time between oil changes provided that the filter is replaced at *each* oil change. (Most newer engines have built-in automotive type filters.) However, operation in dusty areas, cold climates, or infrequent operation will require proportionately more frequent oil changes in spite of the use of a filter. The oil filter element should be replaced after each 50 hours of operation, and it should be cut open to examine material trapped in the filter for evidence of internal engine damage or wear. In new, or recently overhauled engines, some small particles and metal shavings may be present, but these are not necessarily dangerous. Specific evidence of internal damage found in the oil filter justifies further examination of the engine to determine the cause.

Field experience has proven that sometimes mixing brands or types of oil can bring on a sudden increase in oil consumption. This tends to happen at strange airports during refueling when the pilot has not been present during the refueling operation, and a different oil has been added. The line personnel in many cases are not well informed, and generally are at the lowest skill level in the industry. If possible, therefore, the pilot should supervise refueling; if this is not possible, he should *carefully* check what was put in his aircraft.

If your engine needs a quart or two of oil, but the FBO doesn't carry your brand, you might do better not mixing oils. Unless you have a long flight, it might be better to wait until you get home to top off the oil. Better still, think ahead and always keep a spare quart or two of your brand in the airplane.

Fuel

Fuel is very straightforward. Use the grade recommended by the manufacturer. If the exact grade called for is not available, use the next higher grade. Do not use the higher grade all the time.

Fig. 2-4. Avco Lycoming 0-540. (Courtesy Avco Lycoming)

Fig. 2-5. Avco Lycoming 0-320. (Courtesy Avco Lycoming)

There are two reasons for this. It costs more and is simply money down the drain. It does not improve performance nor anything else. In some smaller engines with solid valves, an excessively high octane can cause necking of the valves, which may lead to their breakage and a subsequent engine failure.

Automobile fuels should not be used. The argument has raged for years. The basis for the dispute is that there is no real difference between aviation and automotive powerplants; therefore, they should run just as well on the cheaper automotive fuel as the higher-priced aviation sort. This argument does not hold. Yes, aviation engines and automotive engines are substantially similar in design, but they differ radically in *use.*

The automobile operates at 10 to 15 percent of power at cruise, while the airplane engine is at 60 to 70 percent of its power rating at cruise. This subjects the aviation engine to very high loads at low rpm. This loading subjects the valves to immense heat stress caused by high relative power outputs and the less-than-ideal cooling in an air-cooled engine. Airplanes also operate at high altitudes, and tend to stay on the ground for long periods of time.

These operating demands call for a different blend of fuel than that used in automobiles. Aviation fuel is less volatile and has a lower vapor pressure to avoid vapor lock, which is more critical in an airplane at altitude than in the family car. In addition, automobile fuels have a quicker turnover at the retailer level, so they lack the stability designed into aviation fuels. They turn to shellac readily, which is not desirable—it fouls carburetors and other parts of the fuel system, in addition to causing sticking valves and rings. Road fuel has a high chlorine content. Chlorine is very corrosive in an aviation environment. If this is not enough, road fuel octane ratings are declining every day, making the whole argument moot in any case.

Airframe manufacturers, not to be outdone by Detroit, have taken to hyperbole in describing their aircraft. We are not here to judge the taste of this sort of thing, but one of the current trends is of potential peril to the pilot, and that is the painting of *Turbo-Whoosh* or the like on the side of a turbocharged aircraft. This is frequently just enough to cloud the mind of the line boy and confuse him into thinking it is a turbine aircraft into which should be pumped jet fuel. Airplanes with reciprocating engines do *not* fly well on kerosene.

Even if your aircraft doesn't suffer from one of these paint jobs, the ready availability of turbine fuel at most airports and the transitory nature of the employment of most line personnel is sufficient reason to check your fuel tanks, not only for color, but for smell as well. Turbine fuel smells like kerosene, and is water white in color. A very small amount of red or blue dye left over from the residue of correct fuel in the tanks will color it enough to fool most of us, so smelling is the best test. (Don't overdo it—with the present high levels of ketones in modern aviation fuels, too long a sniff might send you flying without an airplane.) Although aircraft engines do not run well on turbine fuel, once started on gasoline, they will run

Fig. 2-6. Avco Lycoming 0-360, ¾ view. (Courtesy Avco Lycoming)

on turbine if there is a high enough residue of gasoline in the tanks. They will *not* run well enough for takeoff. The attempt has put more than one pilot in the hospital.

Proper Operation

According to Lycoming, improper operation at its worst can put a new engine in the junkyard in 100 hours or less. I suspect this would take a concerted effort, given the quality and longevity of current aviation engines, but somewhere between this time and the recommended TBO is the time you should get from an engine with reasonable care. With exceptional care, TBOs are frequently exceeded. Remember each minute you gain is free time, and, even more satisfying, money in the bank.

Given that few of us are foolish enough to deliberately misuse an engine that we have just bought, the difference between the long TBO pilot and the average or short TBO pilot is how much he knows about engines in general and airplane engines in particular. The pilot should know more about his engine than the automobile driver because the rewards are greater, as are the penalties. Someone who has not the slightest idea of "what makes it go" may have his place on the road, but not in the air. The more you know about what makes an airplane go, the safer you are and the more money you can save. To this end, let's look at some of the things we do, knowingly and unknowingly, and exactly what effect these things have on the engine.

For our purposes it is easiest to divide the operation of the airplane into three areas, and look at what is happening to the engine at each of these periods.

Start and Climbout. A high percentage of engine wear occurs at the start and the first few minutes of operation. If the start is carefully managed, the consequences of this wear can be limited greatly and the life of the engine lengthened commensurately.

Hand-propping to loosen up the engine before starting it with the starter when it hasn't been run in some time, or when the weather is cold, relieves the strain on the starter and gently breaks the initial friction that exists between the rings and cylinder walls and in the bearings. This avoids potentially damaging high initial shock as the starter kicks over the engine.

Hand-propping can be dangerous if done carelessly. Make *sure* the switch is *off*, throttles *closed*, mixture *lean*, and magneto *off*. When hand-propping, treat the engine as if it were going to start. If you haven't propped a Cub or such recently, have a mechanic show you how to do it *safely*—properly done, even if the engine starts by accident, the operator is safe.

Once you know your engine, you should be able to judge the minimum amount of priming necessary to start it under varying weather conditions. This is important, as excess priming washes oil off the cylinder walls, leaving them unprotected at the most critical time in the engine's running cycle. As soon as the engine starts, reduce the throttle to idle stop until oil pressure builds up. Now, advance the throttle to an idle necessary to avoid spark plug fouling and to warm up as quickly as possible. Do not start the engine at a high rpm idle—it isn't getting oil where it needs it most during the initial startup. Oil pressure may take time to build up, because the oil is thick and the oil passages are empty. During this period (and usually for a short period thereafter) it is important to realize there is no oil circulating in the engine. It is running dry, metal against metal, except for whatever small amount of oil is left from the last time the engine was used.

In ordinary weather conditions, by the time you have taxied to the runway and accomplished your run-up the engine is ready for takeoff, but when it is particularly cold sometimes an additional warmup may be necessary to get all the temperatures in the green. Takeoff should not be started with any of the temperatures below normal, even if additional warmup time is necessary.

Run-ups in particularly hot weather should be done facing into the wind, and limit carburetor heat checks to paved runways—with the carburetor heat open, you can suck most impressive quantities of sand and gravel into the engine.

Takeoffs should always employ full power (except turbocharged engines), and power reduction should not be accomplished until the aircraft is able to return to the field and land without power. This is purely for reasons of safety. Catastrophic damage to engines is most likely to occur during power changes, so if there is a rod ready to come out of the crankcase, it will generally choose to do it when power is applied, or surprisingly, when it is *reduced.* An engine may stay together while pulling full power, only to disintegrate when power is reduced.

While maintaining proper climb speed, pay particular attention to your gauges and any leaning you may decide to do. The engine is being used at high power settings during the climbout, and the forward speed is lessened, a combination that tends to heat things up very quickly. If the engine is leaned during the climb, be particularly careful; as a richer mixture runs cooler, and as you deprive the mixture of fuel, it burns hotter.

If you have cowl flaps, make sure they are open; if the numbers get out of hand, richen the mixture and reduce the rate of climb until the temperatures lower, at which point you can resume your climb. Never run the engine at excessive temperatures if it can possibly be avoided; the damage that can be done in even the briefest moments is not worth it.

Cruising. At cruise, the problems we have with excessive heat turn around, and there is frequently a tendency for the engine to run too cold, which is not as damaging as cooking the valves, but tends to be expensive in the long run. After closing the cowl flaps (not only to clean up the airplane, but to raise the cruise temperature to a more efficient range), the engine should be set up for cruise by the book, avoiding higher cruise settings, as they are not only uneconomical in terms of direct fuel expense, they tend to be harder on the engine also. (Lycoming has worked out the following example of fuel savings with a 1977 Cessna 172, in which the cruise setting was reduced from 75 percent to 65 percent at 4000 feet, resulting in an increased range of 40 nautical miles while decreasing true airspeed (TAS) seven knots. Thus an eight percent increase in range was obtained with a six percent reduction in airspeed, increasing flight time *one* percent, a figure so small that it would, in the long run, be eliminated by the reduction in the number of fuel stops.)

The engine should then be leaned as recommended in the manual. If leaning by pulling back the mixture until the engine runs rough is recommended, be sure you know just how to do this. It requires a feel; if you don't have it, have an instructor show you.

When the mixture is leaned in this way, it is not a sign of preignition—it is only the leanest cylinder cutting out intermittently as its mixture gets too lean. The other cylinders are running perfectly normally. At recommended power settings in a normally-aspirated engine, it is not possible to damage the engine by over-leaning. This can happen only at climb and takeoff settings. So don't worry too much about excessive leaning of the engine at cruise. Get it as lean as you can, yet still running smooth.

Descent, Landing and Shutdown. An amazing amount of engine damage can be caused by precipitous letdown. If you want to kill your new engine in under a hundred hours, the next best method after putting a teaspoon of sand down the air intake would be sudden cooling. It will really mess up the insides in short order. Sudden cooling warps valves, causes valve sticking (and thus bent pushrods) and galled guides. In addition, it will break piston rings, crack cylinders and generally play hell with the engine.

When the engine is cruising (65 percent power, properly leaned), it could probably run forever. When the power is suddenly reduced, and the amount of cooling air around the cylinders increased, this equilibrium is destroyed. The valve guides shrink first, reducing clearances, putting pressure on the valve stems, causing the valves to stick and pushrods to bend. The valves cool suddenly, warping, and the cylinder head cools like a hot frying pan in cold water, the hottest parts cracking. The piston rings flutter, breaking and scoring the cylinder walls. This is the worst scenario. Usually things aren't this catastrophic and nothing noticeable happens. We feel we have gotten away with it.

Soon, bad procedures become habits, and we wonder why our engines always need top overhauls out of proportion to their total hours. What happens? Basically the same thing as in the first scenario, but to a lesser degree. The valves do not stick and bend the pushrods; they stick a little, nothing that a little valve guide galling won't cure. They just warp a little bit, so they do not seat properly and tend to wear a little bit more than they should. They don't seat quite as well as they once did, either. This leads to erosion of the valve through improper cooling, cracking of the valve seats and plug bosses, and (slowly and less catastrophically, but just as surely) compression is reduced, and we have a sharply reduced TBO.

If ATC dumps us from high altitudes, there are still precautions we can take to avoid, or minimize, any damage that may occur. Avoid fast descents at cruise rpm and low manifold pressure settings. Lycoming recommends 15 inches or higher, while setting the rpm to the lowest possible cruise setting. This should prevent ring flutter and the consequent broken rings. Letdown speeds should not exceed high cruise speed or approximately 1000 feet per minute. During the letdown, the airplane should be dirtied by dropping the gear and adding some flaps, which will prevent high airspeed, sudden cooling, and plug fouling, while providing the desired good rate of descent. When ATC pressures you to expedite your descent, use any or all of the above methods, and remember your engine, so it won't forget you.

Maintenance

As costs keep spiraling upward, the owner turns more and more toward doing his own maintenance. There is nothing wrong with this, except that some things tend to get missed when, for instance, the owner starts to change his own oil (an area where little expertise is required, and great savings may be realized). When a mechanic changes the oil, he visually inspects the engine for potential trouble. His knowledge of the type lets him know just where to look, and what he is looking at. This is what you miss by doing things yourself—the leaking exhaust gasket, the loose accessory, an incorrect hose routing can be missed by the inexperienced eye. If you are apportioning certain parts of the maintenance of your airplane to yourself, be sure to do this sort of checking at the same time you change your oil or check the plugs, and ask your mechanic what you should be looking for in the way of trouble that can be headed off during routine maintenance.

In any case, follow the manufacturer's recommendations as to oil changes. The manufacturers are not, in spite of what you may have heard to the contrary, in conspiracy with the oil companies to ensure excess use of oil. They recommend oil changes because they have found them to be beneficial to the engine. Oil *does* wear out, and acids, carbon, lead, and other unfilterable impurities build up, rendering old oil not only inferior in its role as a lubricant, but destructive to the engine as well. Saving on maintenance by extending oil filter changes beyond those recommended is false economy of the worst sort. Even if you have to do it yourself, *do it,* particularly if the engine is not flown regularly.

Good maintenance is simply going by the book and checking everything you can see or feel when preflighting the aircraft and airframe. Lycoming feels the average pilot is not sure what he is looking for when he lifts the cowling to do his preflight. To help him they have published the following list. Although I am sure we are all above average, a look at the list would not hurt.

☐ *Fuel*—look for signs of dye which mean leaks.
☐ *Oil*—check for leaks and level.
☐ *Exhaust system*—check for white stains which indicate exhaust leaks between cylinder head and stacks, cracks in the stacks themselves, and condition of heat muffs for cracks and leaks.
☐ *Cowling and baffles*—check for cracks and proper position.
☐ *Air filter*—good tight fit, good condition.
☐ *Anything loose or unattached*—wires, lines, fuel pump, carburetor.

This list is, of course, to be added to your regular check list. When inspecting an engine (or the airplane, for that matter), pay particular attention for signs of trouble. The slight oil leak that has been there since the engine was new is unlikely to give you trouble. If it suddenly gets larger, or there is a new one, that is proper cause for alarm. Have your A&P in for a consultation. Catching minor trouble before it becomes major is an easy way to save money and improve safety.

As you become more aware of your engine, try to avoid tinkeritis, a disease not found in medical books, but common to all shade-tree mechanics. Although it is true that you can't wear out an engine by changing oil, don't adjust things that don't need adjusting. Once you have finished your work, hangar flying is generally more productive and less expensive than tinkering. Take the old mechanic's advice: If it's not broke, don't fix it.

FREQUENCY OF FLIGHT

To the average pilot who doesn't use his airplane extensively for business, the biggest obstacle to achieving maximum TBO is time. An aircraft that is flown daily and otherwise treated properly seldom has much difficulty in reaching or exceeding the manufacturer's TBO. Sitting around airports is as hard on airplanes as it is on pilots.

In these days of high operational costs, most of us feel pretty active if we fly once a week, and our airplanes spend more time on the ground than ever. Lycoming feels that no engine should be run more than six years between overhauls, and specifies that to reach maximum recommended TBO their engines should be run at least 15 hours a month.

What causes an airplane engine to wear out sitting on the ramp? Primarily moisture, which means that your problem is aggravated seriously if you live in a coastal or other high-humidity area. Moisture gets into the engine through the exhaust system, oil breather, and air induction systems, but even if all these orifices were sealed, it would still be present from condensation. There is not too much that can be done about this moisture, other than flying the airplane. Ground run-ups are worse than useless. They aggravate the problem, as it is impossible to get the engine warm enough in a ground run-up to get the moisture and acids out of the oil without developing hot spots in the engine caused by an insufficient flow of cooling air. Prolonged ground run-ups of this sort also tend to cook wiring, seals, and other vulnerable parts of the engine. Airplane engines are simply not meant to be run on the ground. Lycoming states flatly that this sort of run-up is to be avoided.

Moisture, whether from condensation within the engine or

from the orifices of the engine, causes rust in the cylinder walls, especially if the engine has been stored long enough for oil to drain off the cylinder walls. (That's *two days* for engines stored near salt water.) To ameliorate this problem, Lycoming considers the best solution to fly the airplane to get the oil to 165 degrees and keep it there long enough to drive off the moisture and acid accumulated in it. They also recommend 25-hour oil changes for infrequently flown aircraft.

Moisture generally gets at other parts of the aircraft, too. Electrical and ignition parts, as well as avionics, need periodic heating and lubricating to keep them limber and functional, so flying the airplane is probably a good idea. If you can possibly adjust your flying schedule for shorter, but more frequent periods, as evenly spaced as possible, keeping the engine free of water and acid will pay off at overhaul time.

If you can't do this, Lycoming recommends propping the engine frequently to coat the cylinder walls and bearing surfaces with oil. Of course, turning it over with the starter frequently will tend to shorten the life of that component, as well as leave your battery in a perpetual state of discharge, so hand-propping is probably the best recommendation. On the other hand, propping with the starter is easier (and therefore more likely to get done), and more oil pressure can be worked up on the starter than by hand, giving better lubrication of the engine. The starting position of the propeller should be noted when doing this, and it should be stopped 15 degrees off the previous location to avoid stopping the engine at the same location each time.

For periods of enforced inactivity of up to thirty days, Lycoming recommends spraying the interior of each cylinder with two ounces of hot corrosion-preventive oil and turning the engine over with the starter. That will give you a good idea of how seriously Lycoming feels about the matter of unused engines.

In short, there is not much to be done about the frequency-of-flight dilemma. There are no simple answers to the problem; although there are various procedures that may ameliorate the problem to a degree, they are difficult and time-consuming at best. Be aware of the problem, and schedule your flying with it in mind. Changing your flying habits and perhaps finding somebody to fly the airplane with you are the easiest solutions. In any case, try to reach your TBO before six years are up.

Chapter 3

When Is It Worn Out?

If we are willing to put the effort into maintaining and flying an engine properly, with the philosophy that well-cared-for machinery lasts longer and costs less in the long run than mistreated and poorly-cared-for machinery, there remains one important question: When is it worn out?

This question is very important economically, because there is little point to putting extra time and money into preserving an engine if you are the sort of person who will say, "Well, the manufacturer recommends a 2000 hour TBO, so I'll be safe and overhaul at 1500." Manufacturers are loath to talk about the period past the recommended TBO, but they admit that many pilots fly their engines well beyond the recommended TBO.

An automobile engine is worn out when it stops, but we are not allowed this easy solution with our airplane—safety as well as economics is an important aspect of deciding when an engine is worn out. It might even be said that an engine used in an airplane that spends most of its time flying IFR should be considered worn out before one that is used for short hop VFR pleasure work. All of us who fly a lot are well aware that there are enough minor crises waiting for us on every flight. It is not necessary to tempt fate by knowingly flying faulty equipment. There is as much excitement in flying as most of us want as it is.

DETERMINING CONDITION

Having decided that an aircraft engine is worn out before it swallows a valve or blows a jug, it only remains to determine how long prior to these occurrences we should retire the engine. Timing is, as always, very important. Since the ideal from both the safety and the economic standpoint is the same in this case, we may say that the ideal time to retire the engine is after the flight prior to the one during which it will give up the ghost. All we have to know is precisely when it will in fact do so. Lacking a really accurate crystal ball, old-fashioned record keeping is still the best prognosticator and is the best single aid to carrying you to your engine's TBO.

When an engine is new, it is run with straight mineral oil to allow the rings to seat. Sometimes this oil will be consumed at what would be considered an alarming rate in an older engine. That will continue to be the case until the rings seat against the cylinder walls and the ring grooves, at which time the oil consumption will stabilize. Manufacturers use the word *stabilize* to avoid predicting exactly how much oil the engine will use, for frequently there can be large differences in the oil consumption of two otherwise identical engines. The customer who gets the one that requires a quart every two hours may understandably be less happy than the owner whose engine uses a quart every ten hours.

Oil Consumption

What is really important for your record keeping is a good idea of what the *normal* oil consumption of your engine is. If your engine is one of those that use a quart of oil every two hours, don't let yourself be talked into a prophylactic top overhaul after a hangar flying session with a friend who has the same airplane and only uses a quart in five hours. Once you have established the *normal* oil consumption rate for your engine, record quantities used carefully. They will tend to rise gradually over the life of the engine. This gradual increase is normal and no cause for alarm until the consumption becomes excessive. What should really be noted carefully is a sudden *increase* in oil consumption, as this signifies a sudden change, usually for the worse in the internal condition of the engine.

Although a sudden increase in oil consumption can, in the best case, be caused by a leak, generally a seal failure, it is more usually (and more expensively) caused by a serious internal problem, the kind that keeps mechanics in business and pilots poor. The cause is most likely in the cylinders. Cylinder wall scoring and piston ring failure usually are the culprits here. Sometimes, however, this increase is caused by a sloppy bearing allowing so much oil by it (and thence to the cylinder walls) that the rings cannot handle the excess and it is burned in the cylinder.

This symptom is usually accompanied by a lowering of normal oil pressure, particularly when the engine is hot and at idle, although the lowered oil pressure can manifest itself at cruise when the condition gets bad enough. Cylinder wall scoring can frequently be diagnosed with a borescope check. Compression checks and spark plug inspection will usually diagnose worn or broken rings or a general decline in the condition of the rings and cylinder walls.

It is dangerous to operate an engine that is suffering from a sudden increase in oil consumption, as it is usually a reliable indicator of imminent catastrophic failure if the cause is not remedied. If you survive the explosion, an engine with a blown jug or rod out will be difficult and more expensive, if not impossible, to rebuild. Add to this the advantage of having your own A&P do the work rather than a stranger away from home, and not dwelling on the obvious safety aspects, there is more than enough reason not to operate the engine.

The Gauges

The engine gauges in the cockpit are your second line of defense against unexpected shutdowns, as well as being indicators of general condition. Most problems that lead to eventual engine failure are first indicated in the engine gauges, sometimes very briefly, but frequently long before the incident. They also will show engine condition, if a base, or normal, reading is known.

The oil pressure reading is similar in its action to the oil consumption of the engine. The indicated pressure will tend to decline over the life of the engine, just as oil consumption increases. This gradual decline, as long as it stays in the green, is normal and not a cause for worry. But, as with oil consumption, a sudden change in oil pressure (almost invariably downwards), *even if not below the green arc*, is cause for alarm and should be investigated before you operate the engine. To spot a sudden change, you must, of course, know what the normal reading is. You should know your normal pressure under all conditions—cold, idle, climb, cruise, and letdown.

Sudden oil pressure losses are caused by many things, none of them conducive to long engine life. Oil pump failure (partial or total), broken pressure lines, lack of oil, or bearing failure—take your choice, none of them are terribly attractive. Sometimes the gauge may be faulty. It is the first thing to check, because it is the easiest, and cheapest cause of the trouble, although unfortunately not the most common.

The oil temperature and cylinder head temperature gauges are less critical, although equally useless if you do not familiarize yourself with the normal readings under all operating conditions. Excess heat may be caused by cooling air being blocked by a bird's nest or rag, while abnormally cool readings may be caused by damaged baffling, cowl flaps, or other cooling system sheet metal. Operating the engine under these conditions will result in very quick and serious damage that is easily avoidable.

The engine should not be operated under adverse conditions, and that is the primary purpose of these gauges. If the engine normally tends to run hot at climb settings, for instance, the gauge will indicate the condition, but it is up to you to correct it, by either correcting the cause of the overheating, or (as is frequently the case if the condition cannot be corrected), by modifying your operating technique. The engine should never be operated out of the green arc, either too hot or too cold, if you expect to get the most out of it in the long run.

Fuel Consumption

Fuel consumption, like oil consumption and oil pressure, shows a change as the engine gets older. As usual, it's not a change for the better; that is not the way of the world. The increase in fuel use that occurs as the airplane ages is very gradual, and far less noticeable than either oil pressure or oil consumption changes, so it should be monitored carefully. Again, any sudden change should be noted, although it is not the cause for the same sort of alarm that oil system changes are.

Sometimes, increases in fuel consumption indicate ring or cylinder failure, but in these cases they are almost always accompanied by a decrease in oil pressure, as well as increases in oil consumption to help analyze the problem. An increase in fuel consumption more ordinarily would be caused by problems in the fuel system, or ignition or valve misadjustment. A decrease in fuel consumption, particularly at takeoff or climb, can indicate damage caused to the engine by detonation. In this case, the opinion of an A&P should be solicited.

In an engine that is leaned until the weakest cylinder runs rough, an increase in fuel consumption may indicate a failing valve, as the additional fuel will cool the valve and tend to correct the roughness and misfiring associated with a burned valve. If there is a roughness in the engine that can be cured by richening the mixture beyond its normal setting, the valves should be suspect. A top end overhaul is safer and cheaper than a blown jug after the piston has been impaled by a valve.

In addition to the above-mentioned ways of judging an engine's condition, all of which should be well within the average pilot's abilities and skills, there are additional methods which rely on special tools, skills and knowledge. Some of these may be accomplished by most owner/pilots—they are allowed under FAR 43—and the required equipment in most cases is not beyond the pocketbook of the average pilot. There is considerable judgment involved in diagnosing most of these tests. Sometimes a comprehensive knowledge of a particular engine is necessary, so unless you are a highly skilled do-it-yourselfer, they are best left to your A&P.

Oil Changes and Filter Inspections

When the oil is changed, all the screens and filters should be carefully inspected for foreign materials. The oil filter should be cut open and inspected. The material garnered from these sources should be checked by an A&P. Some material in the filter and associated screens is normal, but recognizing what the material is and where it comes from is an A&P's job. The A&P, in fact, will frequently send samples of the material to the manufacturer for analysis. Generally, in a high-time engine the presence of material in these areas, particularly when combined with low oil pressure, high oil consumption, and other symptoms is bad news, but a really good mechanic who knows your airplane is the best judge of what to do. There may be solutions to the problem that are cheaper and simpler than a complete overhaul.

Spectroscopic analysis is a refined extension of the oil filter and screen inspection techniques. While your engine is relatively fresh (it need not be brand new), a sample of oil is taken and spectroscopically analyzed for its impurities. Two or three samples are usually enough to set up a base for the engine, and from there on, a sample from each oil change is analyzed. Any build-up of material shows more quickly and accurately than it does in the filter or screen (the spectroscope indicates contamination invisible to the eye), allowing preventive maintenance to catch the problem before damage can occur.

Many large operators are switching to spectroscopic analysis, but it doesn't seem to have caught on with the average general aviation owner-pilot as of this writing. There is the problem of getting the samples, mailing them, and paying for the service. This is just another complication to airplane ownership, and one with a doubtful cost/benefit ratio.

Compression Checks

Compression checks should be done at regular intervals. They are part of the 100-hour inspection, and they should be done at least that often, even if not required by the FAA. Compression checks, as all other methods of gauging engine condition, should be recorded, as they are most useful in detecting gradual trends in addition to pinpointing failures. A compression check is also useful in determining bad rings, bad valves and seats, blown gaskets, and any number of other problems, in addition to giving a good running signal of the overall internal condition of the engine.

There are two kinds of compression checks; do not confuse them. The standard compression test is no longer generally used by the aviation industry, but it can be done with simple equipment and can be of use to the pilot, since he can perform it himself. The most used compression test today is the differential, or leak-down, check which requires extensive special equipment and is not too practical for the do-it-yourselfer.

The standard check is accomplished with a simple pressure gauge screwed into a spark plug hole. The engine is turned over two or three times on the starter, and the compression noted for each cylinder. This method will spot a weak cylinder, and a repetition of the test with an ounce of oil in each cylinder will tell you whether the rings are the cause of the compression loss (a wet reading), but that is the extent of its usefulness.

The differential or leak-down test has the advantage of exactly pinpointing the source of the compression loss, but requires an air compressor, regulator, a couple of gauges, a limiting orifice (No. 60, if you are considering making a differential gauge) and a spark plug adapter, as well as tubing, fittings, and such.

In use, air (usually at 80 lbs) is pumped from the compressor through the first pressure gauge which measures the compressor pressure, through the limiting orifice to the second gauge and then to the cylinder. The second gauge, therefore, measures the compression maintained by the cylinder. Any leakage in excess of 25 percent pressure differential between the two gauges is considered an indication of trouble. With this test, the exact source of leakage, and thus the problem, can be determined by listening for the hiss of air at the appropriate orifice or area of the engine—an intake valve leak at the carburetor, exhaust valve leak at the exhaust stack, ring leakage or damaged cylinder walls at the oil filler cap, for example.

Spark Plugs

It surprises some pilots, but spark plugs, in addition to their more obvious function of igniting the mixture, are also an important diagnostic tool. This is an area where the skill and experience of an A&P can be of great help, but if you clean and change your own spark plugs, the following list from Lycoming should be of help. If you spot what you think is a potential problem from this list, make note of the cylinder and the position of the plug, and let your mechanic pass on it.

- ☐ Copper run-out and/or lead fouling indicate excessive heat.
- ☐ Black carbon or lead bromide indicate low temperatures and possible over-rich mixture at idle.
- ☐ Oil fouling may indicate unseated or wearing piston rings.
- ☐ In high compression or turbocharged engines, a cracked spark plug porcelain will cause, or may have been caused by, preignition.
- ☐ The normal color of a spark plug electrode is a brownish gray.

OUTSIDE OPINIONS

To paraphrase George Orwell, all engines are created equal, but some are more equal than others. There are engines of the sort we discussed earlier that used less oil than others and run and feel good from beginning to end, requiring few or no repairs in their allotted lifespan. On the other hand, there are those that always seem to be a little bit troublesome, use more oil than desirable, and whose end comes sooner than might be hoped. Your maintenance logbook, bankbook, and memory will tell you which kind you have. Lycoming, while admitting nothing, has the following to say about determining the life of an engine:

"If a powerplant has been basically healthy throughout its life, this would be a favorable factor in continuing to operate it as the engine approached high time. Alternately, if it has required frequent repairs, the engine may not achieve its expected normal life."

Your mechanic's opinion of the condition of your engine should be a major factor in deciding when it has reached the last roundup. Remember, while listening to your mechanic's opinion, to form your own of the mechanic. Some like to save the customer's money; some like to spend it. You should know the mechanic's strengths and weaknesses in this area as you take his opinions into consideration. Your opinions will be somewhere between the extremes and *you* have to be the final arbiter.

Your opinions can also be shaded by the condition of your pocketbook or your desire to make a trip, but don't let yourself be fooled into the school of thought that says, "This shirt can't be dirty, it is the only one I have". Invariably, the dirt shows, and the wearer is the only one unaware of the problem. There will be nobody to help your or take the blame when you suffer the inevitable consequences of pushing an engine beyond its limit; like the next San Francisco earthquake, it is not *if*, but *when*. You know how much you know about engines, and you know how valuable your opinion is; take it for what it is worth.

MAKING THE DECISION

I have not told you precisely how to tell when your engine is worn out. This is far too complex a decision for me to make for you. There are too many factors to be balanced, but your decision cannot be educated unless you consider all the factors. With this in mind, the following recap of the material in this chapter should be used as a checklist when coming to a decision.

- ☐ What is the airplane used for—VFR pleasure trips in flat farm country, or IFR over mountainous terrain at night?
- ☐ Is it a single or a twin? If a twin, how well does it do on one engine, and as important, how well do *you* do on one engine?
- ☐ Oil consumption.
- ☐ Oil pressure.
- ☐ Fuel consumption.
- ☐ Compression test results.
- ☐ Oil inspection; spectroscopic examination results.
- ☐ Maintenance history of the engine.
- ☐ Operation history of the engine.
- ☐ Spark plug inspections.
- ☐ Manufacturer's opinion (recommended TBO; other information).

Chapter 4

Avionics

Flying is no longer the simple business it once was. Even the smallest airplane is almost useless without some avionics aboard, and the average instrument aircraft resembles an avionics platform more than the airplane of old. A complex aircraft can have everything on board that the airliners carry, including radar and an inertial navigation system (Figs. 4-1 through 4-4). All of these things make an airplane more useful for more of the time, and as such have become important in extracting the maximum use out of the large amount of capital invested in the modern light airplane.

The avionics package in a modern light aircraft can easily exceed a third of the aircraft's total cost—the maintenance being commensurate, if transportation, downtime, and other peripheral expenses are included. The initial cost is fixed, but the other expenses aren't. Downtime and transportation can be almost eliminated, while direct repair costs can be reduced with the acquisition and application of a basic knowledge of the avionics system. This does not mean you have to become an electronics technician; in fact you need learn nothing about avionics. The revolutionary thought behind this chapter is the idea that you should know at least as much about your avionics as you do about your airplane.

We all pre-flight our airplanes, getting something out of the experience, depending on our mechanical aptitudes and the care expended. Even the least mechanical of pilots can make engine condition judgements from repeated pre-flights, and can spot the odd missing aileron or wheel, or feel something wrong when the rudder lock is left on the airplane. Why is it then that even the most mechanical pilots, the sort who awe you in the hangar discussing ring clearances and compression ratios, tend to fall ill of the black box syndrome when faced with avionics problems? The black box syndrome is the very expensive tendency to regard all electronics

Fig. 4-1. Everything but radar. (Courtesy Edo-Aire)

Fig. 4-2. What most of us are more than happy to have. (Courtesy Edo-Aire)

as miracles beyond the human ken, understood only by those expensive fellows who inhabit avionics repair facilities.

Why is it that we all seem to be able to use our common sense when faced with a mechanical problem, but hang it in the closet when dealing with an electronic one? If the rudders don't function, we look for ice, a control lock left in place, or perhaps a flashlight or carpet under the rudder pedal. If the radios don't work, we throw up our hands and take the airplane to the radio shop, even if it is 200 miles away.

The average pilot generally has some mechanical knowledge (and this is refreshed during the education necessary to get a pilot's license), but he has no electronic knowledge at all, and none is required to get a license—not even for instrument flying.

This chapter will try to fill that gap, not make you into an electronics "expert" in 30 pages like all those other chapters that you have skipped in all those other books. It will give you just enough background in the avionics of your airplane to make simple repairs and apply your common sense to your avionics problems before spending the next couple of months' mortgage money on unscheduled, unbudgeted repairs. It will also eliminate most of your avionics downtime, and a number of trips to the repairman, a major part of avionics maintenance.

The avionics shop tends to be one of the weakest links in the aircraft maintenance establishment. This is due not only to the technicians, but has a great deal to do with the ever-increasing sophistication and complexity of the equipment itself, as well as the systems used by dealers and manufacturers to obtain factory service and parts.

Often, components show intermittent problems that only the pilot has sufficient knowledge of to troubleshoot and which are impossible to troubleshoot on the bench. The avionics technician is at a loss to discover the problem without a thorough briefing from the pilot, so lacking this, he replaces every out-of-tolerance part in the set, or sends it back to the factory, where they must necessarily

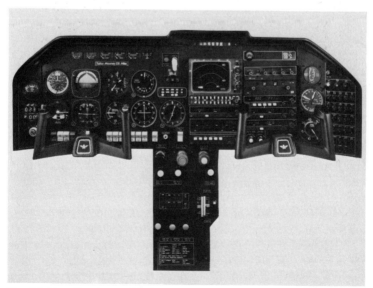

Fig. 4-3. The ultimate panel in the ultimate airplane, 1981. (Courtesy Mooney)

Fig. 4-4. Edo-Aire's RT-553 nav-com. (Courtesy Edo-Aire)

take the same approach. This approach frequently does more harm than good. In any case, this method too often has no noticeable effect on the actual fault. It is this sort of difficulty that costs the owner/pilot the most, and has given the avionics industry a disproportionate part of its reputation for inefficiency. Oddly, it is this sort of problem that is most amenable to pilot analysis.

A great help in avoiding problems of ignorance (both yours and the avionics technician's) is to obtain the manuals and service bulletins for your equipment, especially any of it that may be old or a little offbeat. Read it yourself (it is unlikely to hurt, and it may even help), and keep it in the airplane, particularly on a trip away from home. It may allow a repairman unfamiliar with your equipment to make a repair he wouldn't even consider looking at without it.

Of course, as in all other things having to do with your airplane, make sure everything is working *before* you leave. Repairs away from home are almost always more expensive than when the work is done at home. Part of this is due to the "We'll never see him again" attitude prevalent at many repair facilities, and part to the rush nature of repairs while on a trip, not to speak of motel rooms, taxis, and so forth concerned with a breakdown in a strange place.

The mechanical parts of avionics tend to wear out too, and the life of these can be immeasurably increased by thoughtful care. Operate electronics with the respect their cost deserves. Turn knobs gently, and get in the habit of taking the shortest route to your desired frequency or code. *Never* leave them on when you turn off the master switch. This is an open invitation to having them on when turning on the master, the worst single thing you can do to your avionics. Modern installations have master switches for the avionics—something that may be worth adding to an older airplane if you have trouble remembering to switch off the avionics before switching the master on.

PRE-FLIGHTING AVIONICS

One of the many good reasons that there is no avionics checklist in your airplane is the diversity of avionics installations, which precludes a ready-made list. This in no way implies that one is not necessary, but it does mean that you have to make your own, keeping in mind the peculiarities of your avionics, installation, and airplane.

Your avionics should be pre-flighted along with the rest of the airplane. The reliance placed on the avionics by the modern pilot leaves no room for the "I guess the tanks are full" approach to flying. In addition to the obvious benefits that will accrue from knowing your avionics better, the monetary savings should not be forgotten.

Fig. 4-5. Installing avionics on the Mooney production line. The best argument is for a factory installation. (Courtesy Mooney)

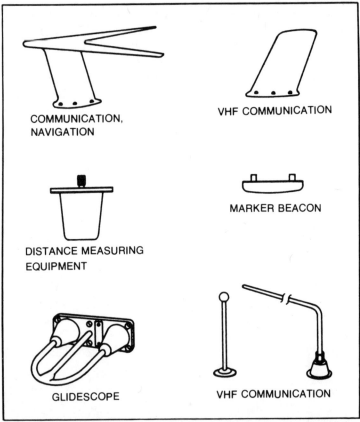

COMMUNICATION,
NAVIGATION

VHF COMMUNICATION

DISTANCE MEASURING
EQUIPMENT

MARKER BEACON

GLIDESCOPE

VHF COMMUNICATION

Fig. 4-6. Typical modern antennas and their functions. (Courtesy Edo-Aire)

The more you know about avionics, the more you will save in repairs.

The first of the following lists is an adaptation of the FAA's *Pre-Flighting Your Avionics* publication, and a good starting point from which to construct an everyday checklist for your airplane and installation. It will get you off the ground knowing what is, and is not, working, while keeping you from blasting your avionics with the initial surge of voltage from your master switch, and the like.

The second list was compiled by Jack Knott, Field Service Engineer for Edo-Aire, to help owner/pilots isolate problems in their electronics. Completely run through as a periodic checklist with the resulting figure recorded in a log, it will allow you to establish normal tolerances for your avionics, just as you establish normal oil and fuel figures for the engine.

Run through and carefully filled out, this list will also allow you to describe to a technician what is wrong with the equipment in terms that will mean something to him. It is part and parcel of the system this chapter recommends for saving time, money, and inconvenience in the care and maintenance of your avionics. You do the time-consuming and expensive detective work; let your repairman do the actual repair and replacement.

The Daily Pre-Flight List

Walk-Around

Check all antennas for physical condition, cracks, oil or dirt, proper mounting and damage.

- ☐ Comm or nav comm.
- ☐ VOR.
- ☐ Transponder.
- ☐ Marker beacon.
- ☐ Glideslope.
- ☐ ADF sense.
- ☐ ADF loop.
- ☐ DME.
- ☐ Radar.

In Aircraft

- ☐ Make sure FCC license for station is on board.

Before Starting Engine

- ☐ All avionics OFF.

After Starting Engine

- ☐ All avionics ON.

Communications Radios

- ☐ Set proper frequencies, memory, if any.
- ☐ Audio switches, select headset or speakers.
- ☐ Squelch control, adjust.
- ☐ Volume, adjust.
- ☐ Transmitter, select desired transmitter.
- ☐ Radio check, *both* transmitters and receivers.

VOR Radios

- ☐ Proper frequencies set, memory, if any.
- ☐ Flags.
- ☐ Identification.
- ☐ Accuracy, when possible from VOT.

DME

- ☐ Readout, when possible.
- ☐ Identification.

ILS

- ☐ Frequency set.
- ☐ Flags, localizer, and glideslope.
- ☐ Identification.

Marker Beacon

- ☐ Lights, test at both low and high frequency.
- ☐ Audio, on, then adjust volume, if possible.

Transponder

- ☐ Code, set.
- ☐ Switch set standby.
- ☐ Circuitry, if test switch is provided.

ADF

- ☐ Frequency, set.
- ☐ Identification, check.
- ☐ Select mode, and confirm accuracy, if possible.

Before Takeoff

- ☐ Transponder on.

At Cruise

- ☐ Check items on list that were not possible to check on the ground.

Edo-Aire's Periodic and Troubleshooting Checklists

In case of trouble with your avionics, carefully fill out the General Failure List, then turn to the list for individual components. (Only the more complex or popular components are represented.) Work through each question as soon as possible, while your memory is still fresh. Thoroughly acquaint yourself with the material in the text, try the recommended solutions, if applicable, then take the problem to your avionics repair shop, with your notes in hand, so you can describe the problem to him. By this time, you should have a pretty good idea of just where the problem is; if you don't, the repairman should be able to figure it out pretty closely with the material you have supplied.

Avionics General Failure Report

- ☐ How long into flight did problem occur?
- ☐ On what frequency does a frequency change correct or

modify the problem?
- [ ] What were the flight conditions, weather, temperature, precipitation, ice, etc?
- [ ] Route of flight.
- [ ] Detail and complete description of problem.

Communications Receiver

- [ ] Sensitivity. How far away can you receive a given station? Has this capability diminished?
- [ ] Audio, quality, loudness.
- [ ] Squelch. Check operation. If automatic, it can malfunction, cutting out all audio.
- [ ] Frequency. Does the set operate on all frequencies? If not, which ones?
- [ ] Interference. What kind?
- [ ] Problem occurs when unit is:
    —hot.
    —cold.
    —vibrated.
    —on for more than an hour.
    —other.

Communications Transmitter

- [ ] Keying.
- [ ] Audio.
- [ ] Frequency.
- [ ] Microphone, type used, remote switch?
- [ ] Range.

Navigation Receiver

- [ ] Sensitivity (range). See manual for usual range. You should know yours; a change can be important.
- [ ] Audio, quality. It should be clear, clean. Check by listening to an ADIS. Compare the quality with the comm receiver. It should be comparable.
- [ ] Ident filter. This filter allows you to listen to voice by filtering the nav information. If it doesn't filter, it is difficult or impossible to hear the voice. If it works too well, it can disturb the nav information.
- [ ] Frequency. Do all frequencies function?
- [ ] Interference. Try to pinpoint.
- [ ] VOR and OBS course width. See manual for usual numbers.
- [ ] Right-left needle stability. If unstable, at what range?

- [ ] Right-left needle sticky, etc. Check by rotating OBS.
- [ ] VOR to-from flag operation. Sticky or other defects?
- [ ] LOC centering.
- [ ] LOC sensitivity, course width.
- [ ] Glideslope centering.
- [ ] Glideslope sensitivity.
- [ ] Glideslope stability.
- [ ] Glideslope needle movement.
- [ ] Glideslope flag operation.

Automatic Direction Finder

As a check, go through list. For troubleshooting, start with engine off, proceed to engine on, alternator or generator, on/off to pinpoint source of interference problem.

- [ ] Sensitivity in REC position, LF-BC.
- [ ] Audio, quality, interference.
- [ ] Tuning frequency, display accuracy.
- [ ] Sensitivity, in ADF position LF-BC.
- [ ] Audio, in ADF position.
- [ ] Frequency at which problems may occur.
- [ ] Indicator accuracy.
- [ ] Indicator stability, if oscillating, fast or slow.
- [ ] Test function.
- [ ] Needle, does it operate smoothly through 360 degrees?

KEEP THE WEATHER OUT

Water and heat are the main culprits in increased maintenance costs and the early junking of electronics, although they may both be avoided by proper installation, care, and use.

When you have avionics installed, insist on a properly-installed cooling kit, and make sure that it doesn't leak water. With recent existing installations, it is frequently possible to install a cooling kit. If it is possible, do it. Older installations in older airplanes should be cooled as well as possible, even to the point of moving around the components for better ventilation, if feasible.

Once you have a cooling kit installed, or have carefully ventilated your present installation, do not negate the effect of your expensively engineered and installed kit by covering the vents with a $1.85 chart. If you plan to use your instrument panel shelf as a chart case, you might as well save yourself the expense of the cooling kit and resign yourself to high avionics maintenance expenses.

When installing a new radio, never install it *above* an old one. Old radios with their vacuum tubes and other heat-producing components can fry a new radio very quickly. The new avionics should always go on the bottom, even if it offends your sense of aesthetics and order.

If the airplane has been sitting on the ramp in the Arizona desert on a hot summer's day and you can fry the proverbial egg on the instrument panel, open the doors and let the cockpit cool off to ambient temperature before switching on the avionics. If you have a blower on your cooling kit, turn it on to speed the job of lowering the avionics temperature. Avionics will survive very high temperatures without damage when they are shut down, but turn them on when super-heated and all hell breaks loose. An incredible amount of damage can be done with the initial voltage surge, so keep them switched off until they are as cool as possible under the circumstances.

Anything you can do to keep the avionics cool while on the ground is helpful. Cockpit covers are useful in this regard, but tend to be too much bother for short-term use. I find it best to leave the doors open while the aircraft is being fueled or you are having lunch. (I have never had anything stolen, but I wouldn't leave my wallet on the front seat, either.)

In many types of aircraft, the list of special equipment for wet-weather instrument flying includes not only the usual stuff, but also a sponge for sopping up the water that cascades into the airplane from the windscreen seal and other weak points of the design. This situation is not just a bad joke perpetrated on pilots by some manufacturers, but it can be dangerous and expensive as well. Frequently, the visible water coming in around the windscreen represents only a small percentage of the total—the rest is cascading in rivulets down the *inside* of the instrument panel, washing the insides of the instruments and turning your avionics into a tropical paradise on its way to the bilge, where it does its best to render useless any antenna unfortunate enough to the mounted in the belly.

Even if your avionics and instruments survive this periodic deluge, it doesn't do them any good. Frequent bathing is not on their maintenance schedule. A superhuman effort should be made to stop leaks (particularly in the windshield and associated areas) in any airplane, old or new, as they are dangerous to the avionics and instruments, not to speak of the dispiriting effect they are likely to have on both pilot and passengers while flying in trying conditions.

When you finish correcting the airplane's design errors and

have the cabin watertight, make sure the belly is properly drained. Some water seems sure to find its way in here, and it is not really designed to function as is the bilge in a boat. Water, even in very small quantities, can do an awful lot of damage in the long run. Keeping water out of your airplane has the additional benefit of eliminating the terminal jungle rot that affects both the odor and resale value of so many aircraft. One would think it would be possible to seal up the interior of a $60,000 airplane as well as a $6000 car.

NEW INSTALLATIONS

Jack Knott, Field Service Engineer for Edo-Aire, reports that 60 percent of Edo's troubles with new avionics are in the installation. In an area where absolute reliability is essential, it seems to be an elusive goal.

The varying quality of available avionics installation and service creates a thorny problem for the pilot in choosing an avionics shop to do his installation and repair. An airplane equipped with the avionics described in this chapter has on board seven receivers and five transmitters operating through as many as eight antennas. The avionics fight for space on the panel and cooling behind the panel. The wiring fights for space and priority all the way to the antennas, and the antennas themselves compete for valuable space on the fuselage with other add-ons and the aircraft components themselves.

Every one of these components has the potential to foul one another's signals, rendering their information useless and communication impossible, and generally make flying a dangerous nuisance. They can also be affected by signals generated by other components in the aircraft, and the incorrect routing of their power and other cables and antennas can pick up these signals, as well as foul controls and wear out cables by abrasion.

Avionics installation is exacting, complex, and important. There are very few technicians fully capable of installing *all* the avionics in *one* type of airplane—fewer yet who can install them in a large variety. Thus, you will probably do best, if you are buying a new airplane, to have the avionics installed at the factory (Fig. 4-5). If you are buying the brand of airplane that sells only its own avionics, go to a shop that specializes in avionics installations in that airplane as well as the brand of avionics you want. Do not make price your *only* criteria of choice. Check with other pilots. Inspect the installer's work. Even the most inept of us can spot the main feature of a good installation: neatness. All the wires should be wrapped in incredibly neat bundles; in short, the installation should look like the pictures you have seen of the wiring in the Apollo space capsule.

Be sure you settle the terms of the installation before signing anything or paying for it. Make it clear that there is to be *no* noise *anywhere,* under *any* circumstances, and that performance is to be up to par and everything is to function perfectly *before* you accept delivery. It is better for the dealer to have problems with his avionics rather than you with yours, and they are *his* until you pay for them.

If you have a continuing problem after installation, remember the opening sentence of this section and contemplate the logic of having the installer, who most probably made the mistake to begin with, try to correct it. In the case of an installation that *never* has worked right, stop calling the dealer; call a lawyer. There are many new consumer protection laws that cover merchandise that doesn't work as intended. In many cases, acceptance can be revoked; that is, the contract can be abrogated and the manufacturer (or dealer) made to take the merchandise back.

WHAT CAN I DO?

Unlike most of the maintenance of your airplane that is regulated solely by the FAA, the radios are additionally regulated by the Federal Communications Commission (FCC), which is why you have to have a license aboard your aircraft for the radio station.

The FCC is very clear in delineating what you, as an operator, may and may not do to your radios. You may *not* work on a transmitter without a license from the FCC authorizing you to do so. The rest of the avionics equipment on board is your business. The FCC is not interested in protecting you from your own folly, only in protecting their airwaves from illegal and offensive signals that may be transmitted by misaligned or misrepaired transmitters.

The FAA on the other hand, is more benevolent, intent on protecting you from your own ineptitude, and they allow you to do all but "major" repairs. This, to be sure, allows considerable latitude, but is a little bit fussier than the FCC. Interestingly, the FAA does not define their idea of "major."

Basically, it is probably safe to say that if you are reading this chapter, and getting something from it, you have no business in any of the "black boxes." The troubleshooting methods described in this chapter should keep you out of trouble, and are unlikely to be construed as "major" by even the most hard-nosed FAA bureaucrat.

Antennas

Antennas and their associated connectors cause about 40 percent of the faults in older installations (Fig. 4-6). Most of these can be prevented, discovered, or cured by the average pilot using the same techniques employed by an avionics repairman, but with the advantage that you know more about the problem and its history than the technician. In addition, you probably work cheaper. All you have to learn is which antenna is which. Antenna design has changed radically over the years in an attempt to correct faults in earlier types, which sometimes makes them hard to recognize. The newest antennas are covered in plastic to minimize weather damage, vibration, and ice, and are designed to minimize installation error and need for adjustment. If you have one on your airplane that you don't recognize, check with your A&P or avionics technician. You really should make the effort to know which is which. After all, you know what the ailerons and rudder do, don't you?

A visual inspection of all antennas should be made a part of your pre-flight inspection, even when flying around the pattern. Certain antennas are susceptible to ground damage, and a familiarity with these vital appendages will allow you to notice when trouble is on the way, rather than after it arrives. Water dripping from mounting points on the belly of the fuselage, corrosion of the fasteners or mounts, rust, cracked or broken insulation, and other damage will be readily apparent once you know from experience how things should be (Figs. 4-7 through 4-11).

Be particularly careful after a flight in bad conditions or ice. Ice in particular is very hard on navigation antennas, and it is not rare to find one or both sides of the antenna bent or broken after flying through an apparently insignificant amount of ice. The modern plastic-covered antennas are less likely to suffer damage, but when they do, it can be very difficult to spot, as the antenna is a very fine piece of wire buried in the plastic. This wire can break without much if any external evidence. With half or all of the navigation antenna gone, you hardly want to continue a flight in the sort of conditions that would tend to break it.

The same sort of problem can occur with the sense portion of the ADF antenna. If the installation is bad, or the insulation is cracked or broken, it becomes nearly useless in bad weather. In these modern, sophisticated days, the ADF is used less frequently than it used to be, but when it is required as a backup to the other navigation radios, or as the primary source of navigation information on an approach, a high degree of accuracy and dependability is required, particularly in bad weather.

The little DME and transponder antennas also seem to suffer more than their share of trouble—most of it caused by the hands of over-enthusiastic ground personnel intent on getting the belly and all protuberances clean and oil-free. Excessive zeal frequently results in broken or bent antennas.

All antenna installations, new or old, are subject to corrosion and vibration. Vibration damage is frequently hidden on modern antennas, as is corrosion. A broken internal wire will evidence itself

Fig. 4-7. ADF Sense antenna with cracked and broken insulation. Water here will make the ADF useless. (Courtesy Edo-Aire)

Fig. 4-8. A navigation antenna, and another faulty ADF sense. (Courtesy Edo-Aire)

in faulty operation, while corrosion damage will frequently leave weeping rust marks from gaskets or faulty or improper mounting hardware or fasteners. These fasteners are supposed to be stainless steel; if they aren't, they corrode, resulting in a faulty ground connection. Modern antennas resist corrosion, but they must be installed correctly and with the proper hardware or they lose their advantage. Cracks in insulation should be looked for, as they defeat the purpose of the insulation and will eventually cause a deterioration of performance or a failure in your equipment—usually in bad weather, just when you need it most. Remember that all belly-mounted antennas are subject to water damage caused by water in the belly of the aircraft, so try to keep it out.

Static wicks are not antennas, but they should be checked for tight connections and overall condition. If there is any doubt, have them replaced. This is cheap insurance.

Communications Transmitter-Receiver

The very high frequency (VHF) communications transmitter-receiver (transceiver) operates on 720 channels reserved by the FCC for aviation use. The receiver portion of the set is an amplitude modulation (AM) receiver, much like the one in your automobile,

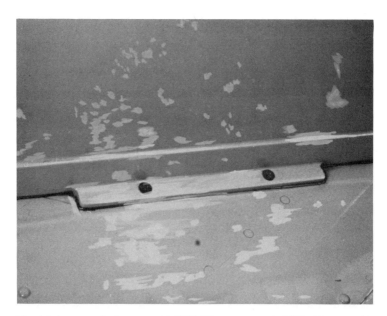

Fig. 4-9. Improper fasteners on this VHF. They are corroded. What happened to the ones the antenna manufacturer furnished? (Courtesy Edo-Aire)

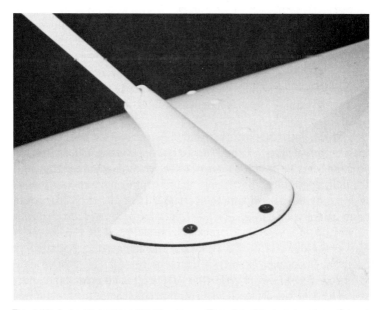

Fig. 4-10. A plastic enclosed VHF antenna. The wire often breaks where the rod goes into the base, where it cannot be seen. (Courtesy Edo-Aire)

Fig. 4-11. The DME antenna, clean for a change. (Courtesy Edo-Aire)

but tuned by crystals for greater accuracy. It has an antenna to pick up the radio frequency signals from the air, which it amplifies and converts to audio frequencies which are made available to the listener by means of a loudspeaker or headphones. The transmitter portion of the set is simply a reverse of the receiver. It takes the pilot's voice (audio frequency) through the microphone, converts and amplifies the signal, and transmits the radio frequency signal over the crystal selected frequency through the antenna (Figs. 4-12, 4-13).

Since both the receiver and the transmitter use the same frequency and antenna, they cannot both operate at the same time, so there must be a switch to turn off the receiver when the transmitter is in operation. This is only the first of many complications of what is, essentially, a simple system. If the aircraft has two communications transceivers (comms), they are generally run through a master panel so one or another may be selected by the operator. This panel also switches the microphone and speaker to match. In addition, this panel frequently contains the marker beacon indicators as well as switches for the ADF audio and other functions.

The first line of attack on all radio problems and malfunctions is effective isolation of the problem. Use your common sense and the lists in the previous section of this chapter. Is the problem in one or both radios? Is it on one or both sides of the radio—that is, if there is a problem with the receiver, is it also evident in the transmitter? If there is noise, what kind is it—static, electronic, or electrical? If it is an electrical noise, does it vary its tone with an engine rpm change? If it does, it is probably caused by the ignition or alternator/generator. Other noises might occur when the gear or flaps are activated. Electrically-operated flight instruments sometimes cause interference—any electrical motor or device on the airplane is a potential source of interference. Electronic hums or buzzes keep the same tone independent of the operation of the engine or accessory motors, and are generally in the wiring or interior of the set.

A large proportion of avionics troubles are in the antenna and associated wiring circuits and connectors (Figs. 4-14 through 4-16). Not only do problems in this area represent the largest single group, but they are usually amenable to diagnosis and repair by the owner/pilot, so in troubleshooting, we start here first, for if the problem is here it is usually cheap to repair. In addition, faults in the antenna and associated circuits have to be eliminated before we can pinpoint trouble within a set.

The antennas of the afflicted set should first be visually inspected as discussed in the antenna section of this chapter. If the

Fig. 4-12. Edo-Aire's RT-551, R-552 nav-com. (Courtesy Edo-Aire)

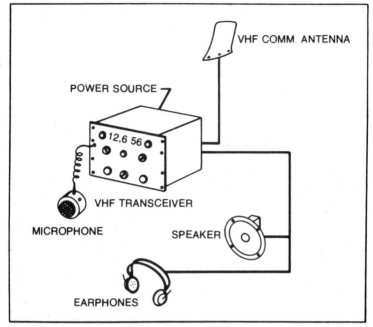

Fig. 4-13. VHF system. (Courtesy FAA)

problem is static or any other sort of intermittent noise, tap the external portion of the antenna assembly with a broom handle *(gently)* or similar appliance with the set turned on and try to induce the problem. If the fault is in the transmitter portion, broadcast to a receiver near enough to your transmitter to allow you to hear when the problem occurs. A friend on another radio will often suffice. If this test is negative, try pulling, twisting, turning, and tugging (very gently) the connectors and antenna leads, both at the antenna and at the set, again trying to induce the fault. Open up the connectors and make sure the connectors are internally clean, dry, and corrosion-free. Move the antennas leads around and visually inspect for abrasions and wires coming loose from connectors.

All this wiring is generally readily accessible through zippers in the headliner and easily removable panels and carpets. With modern installations in which the radio is installed in a rack, push or tap on the radio to pick up troubles in the connectors in the rack. If there is an electronic hum, move the leads around and listen for a change. Frequently these hums and other stray noises are picked up by antenna and other leads running close to or parallel with other wires, picking up interference from them in spite of their shielding.

Look for faults in the connectors, leads, or shielding, as any of these can leak noise into the system.

If you want to shift a lead out of the way of a source of interference, be very careful that it doesn't interfere electronically with a new lead or cause interference in another wire. Also, be sure that in its new location it doesn't foul the control system or rub against a moving or fixed part of the aircraft.

Sometimes, noises are caused by problems in another part of the airplane. If you get a lot of static when an electric motor is turned on, there could be a problem with the motor, or a radio interference suppression capacitor in the motor could be faulty, or there could be a problem with the radio itself.

Fig. 4-14. "Tap gently with a broomstick;" a com antenna. (Courtesy Edo-Aire)

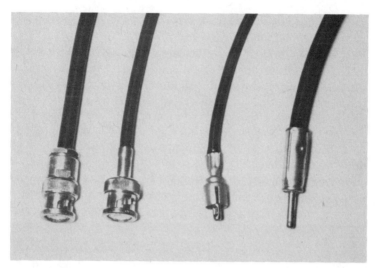

Fig. 4-15. Connectors used for avionics installations. (Courtesy Edo-Aire)

When a system is new and the avionics are functioning at their peak capacity and the shielding is new, noises and other interference have a hard time finding their way into the system. As the avionics and the installation grow older, the performance declines and all sorts of problems that weren't evident when the installation was new show up.

Watch out for problems that aren't problems at all. VHF communications sets don't work well at lower altitudes, so if you need communications, keep your altitude up. If you are lucky enough to have a comm installation with two antennas (to avoid blanking transmissions at certain aircraft attitudes), make sure you know which radio is attached to the belly antenna. That is the radio *not* to use when calling ground control or in general communications from the ground. VHF is line of sight, and an antenna ten inches above the ground and under an airplane can't see much!

Even if you do not feel up to repairing troubles in these areas yourself, you can save a lot of the avionics technician's time by diagnosing them yourself and telling the repairman exactly what the problem is, particularly because these problems, while simple, are the ones that don't show up in bench testing.

Microphones, Headsets, and Speakers

Microphones have their own set of troubles, all of which are owner-repairable. Microphones come in two flavors, dynamic and carbon. The dynamic has superior fidelity, although it is not as rugged as the carbon. Although the dynamic microphone should be more trouble-prone, in practice it doesn't seem to be, and has almost supplanted the carbon microphone.

Carbon microphones are simple, and have only one problem that is unique to them: they rely on loose carbon granules in the microphone to transmit sound. In time, these granules (especially with the vibrations common in aviation use) pack together causing deterioration in the already-poor quality of the transmission. In an emergency, a sharp rap against a hard surface will frequently ameliorate the problem (or destroy the microphone), but the best cure is a periodic (2-3 year) replacement of the carbon granules in the microphone. If you have an old one for which the replacement cartridge is not available, get one that has the replacement cartridge feature—or better yet, a dynamic microphone (Fig. 4-17).

Dynamic microphones generally have electronic circuitry in the microphone, which makes it more fragile than the carbon sort and could (but doesn't seem to) cause trouble. On the whole, the dynamic microphone is almost as trouble-free as the carbon sort, and it requires no periodic maintenance or replacement of cartridges.

All microphones should be held as close to the lips as possible to eliminate noise (of which there is plenty in an airplane) and feedback, the high-pitched squeal to which the dynamic microphone is especially prone. The main reason why boom microphones tend to be disappointing in their performance is that they are necessarily held too far away from the mouth, allowing noise in (Fig. 4-18).

Fig. 4-16. A shielded lead. Note the shield (the braided portion) and the center lead. A shield must be properly grounded and unbroken to do its job. The center wire is very small and can be broken easily, particularly where the wire enters connectors. (Courtesy Edo-Aire)

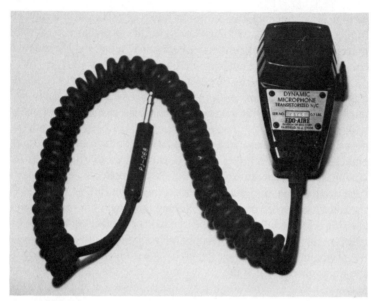

Fig. 4-17. A dynamic microphone. (Courtesy Edo-Aire)

Wires are sometimes a problem. Microphones tend to get banged about and the wires wear out where they go into the jack, as well as where they go into the microphone, even when heavily reinforced (Figs. 4-19, 4-20). Wiggling, pulling, twisting, and otherwise abusing (gently) these connections while the transmitter is keyed will usually reveal the exact source of this sort of trouble, and both the jack and the microphone can usually be easily disassembled (and reassembled) to inspect for and repair this sort of internal malady.

If you have a two-microphone installation, you should bear in mind that a problem in one can frequently appear to be a problem in the other, or cause both to malfunction. The first thing to do when you suspect a problem of this nature is to disconnect both and try each one separately, making sure it is firmly plugged in. Sometimes, if the jack is pulled a little out of its socket, it will short, disabling itself and sometimes the other microphone, so it is important that they both be firmly plugged in.

Don't exchange microphones between airplanes. Dynamic microphones have differing outputs, and the gain control on the transmitter must be adjusted for each microphone. Swapping microphones can cause transmitter overmodulation which can make your transmission unintelligible, damage your transmitter, or even cause erroneous navigation readings. Turning up the gain adjustment of the transmitter is not the easy way to increase the power of your transmitter—it is there to tune the microphone, and should be left to your avionics technician to adjust. Remember the FCC—this is not a do-it-yourself operation.

Microphone buttons tend to stick, sometimes externally when the button rubs against the casing, and sometimes internally, when the problem can be less obvious. The problem is aggravated when there is a button on the control yoke. Make sure all your microphone buttons are working properly. A jammed button at the wrong time can cause a lot of grief, particularly if you don't know it is jammed.

Headsets (earphones) have few problems that are uniquely their own. Their main problem is shared with microphones—wear and tear on the wiring. This can be checked and repaired just as you would the microphone. A problem with a headset should be checked by substitution (substituting the speaker for the headset to ascertain that the problem is indeed in the headset).

Speakers do fail, particularly in the coil which activates the paper cone, but speakers also tend to deteriorate over the years and should be checked from time to time, particularly if you notice a general deterioration in quality. It is likely to be caused by the

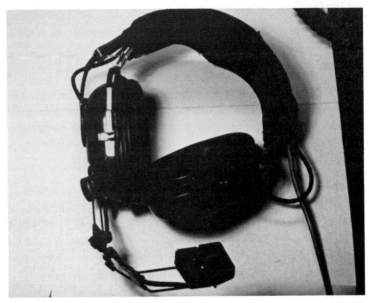

Fig. 4-18. A headset with microphone. Keep the microphone as close to the lips as possible. (Courtesy Edo-Aire)

Fig. 4-19. Where wear occurs. (Courtesy Edo-Aire)

deterioration of the paper cone, frequently evidenced by cracks or tears in the cone. Speakers are cheap to replace, but make sure the impedence of your new speakers matches the old one. (You don't have to know what impedence is, just make sure it matches.)

VOR Receiver

Frequently, the pilot who manages quite well in troubleshooting his communications set will throw in the towel when faced with the complexity of his navigation receivers. Most pilots feel that these sets are too complex to yield to simple methods, and are all too aware of the reliance they place on their VOR receivers when flying under instrument conditions. These days, in fact, many of us would be lost without our VOR receivers even on the clearest days. Their accuracy and dependability have to be maintained at the very highest levels.

The VOR receiver, is in fact, simpler than the comm set, and the reliability and accuracy of the set is only helped by the pilot who is fully aware of the way it functions.

The VOR system operates on FCC-assigned frequencies between 108 and 118 MHz. The receiver is basically the same as the VHF receiver. The audio is taken directly from the receiver, put through a filter to reduce the identification tone so the voice transmission may be heard over it, amplified, and sent through a switch to the loudspeaker or headphones. The same signal is also routed to the processing circuitry where it drives the OBS, left-right needle, and to-from indicator, as well as the function flag.

Over the years, a great deal of time and money has been spent on avionics problems that didn't exist in the first place. Knowing exactly what the VOR receiver will and will not do eliminates many false starts at the avionics repair shop.

Always remember that terrain can cause errors in VOR readings. The worst of these will usually be noted in the charts, but not all are, and some noticeable ones are likely in certain kinds of terrain, although they might not be serious enough to mark on the charts. Errors can also be caused by changes in terrain or buildings near the VOR transmitter, while seasonal errors are often induced by snow or ice near the VOR. Doppler beacons are now being used where terrain precludes the use of the ordinary kind. These work extraordinarily well in difficult terrain, but they can give a slightly different bearing than the conventional sort of station, and this should always be expected.

Another problem that at first may appear to be terrain-related is propeller modulation of the signal. If the attitude of the aircraft is

Fig. 4-20. The wiring inside the jack can also cause problems. (Courtesy Edo-Aire)

such that the VOR signal has to pass through the propeller to reach the antenna, it is possible that it may be modulated to a slight degree by the propeller, causing a reading error or a needle swing at the VOR head. If you suspect this sort of problem, try changing the propeller rpm, and look for a change or improvement.

An entirely different problem, but one in which a change of rpm also can cause a change in the readout, may be caused by fluctuations in the bus voltage of your airplane's electrical system and/or an intolerance for this sort of fluctuation in the VOR receiver. An error of a degree or so is to be expected between alternator on and alternator off readings in a normal VOR receiver. Anything more than this (or any change in the usual error) should be adjusted out or repaired, as it is a sign of trouble which is only going to get worse. We all know that a sudden 15-degree error in navigation heading is not what we need when the alternator has suddenly failed and our workload has increased to near intolerable levels already.

Very often this sort of large navigation error noticed on throttling down the engine (as in landing) is the first indication of a major defect in the electrical system and has nothing to do with the VOR receiver. If you get this sort of reading, check it out *immediately*. The source of the problem will be a substantial voltage drop that brings the bus voltage down well below its usual 11-14 (22-28) volts. Anything that causes this situation should be corrected immediately. It can be a failing battery, short circuit, or any number of unpleasant things. Frequently, a straying VOR readout is the first sign of an alternator or battery failure, although it is usually not noticed until you are well off course and without effective electric power.

Atmospherics are a broad group of conditions that disrupt radio transmission and reception. They are caused by sunspots, weather, changes in the ionosphere, and other unknown causes. They are readily noticed on voice communications, but less so in navigation receivers. If you are having trouble with the needles, particularly at night, you can check out for atmospherics by listening to the audio portion of the navigation signal. Usually, if atmospherics are the fault, you will hear intermittent reception of two stations (or more). The only cure is to try another station, switch to the ADF (although it is even more likely to be affected) or fly another day. Atmospherics that affect the VOR will not necessarily be evident on the VHF set—sometimes they are very specific as to frequency.

VOR receivers also have a small error due to distance. When crowing to your flying buddies about the great distances at which you can pick up a station, remember that this sort of long-range reception usually includes a small error. Sometimes, there will be a noticeable difference in closer bearings—at thirty miles, or even when coming up on the beacon at one or two miles. As long as these errors are small and stable, they can be considered tolerable. But as with other errors noted, if they get out of hand or obtrusive in any way, they should be adjusted or repaired out of the system. Any sudden changes in these slight errors should be viewed with intense suspicion, as they are invariably harbingers of component failure.

The fact that the VOR receiver and the communication receiver are basically the same (at least through the main part of the receiver) make troubleshooting similar. The differences and complexities of the VOR receiver start at the converter where the signal is dismantled and channeled to the OBS and the left-right needle to give the pilot useful navigation information (Fig. 4-21).

The first place to check for trouble is in the antenna and connector system, just as you do with the comm set (Figs. 4-22 through 4-24). When checking the antenna, tune in a strong, nearby station and turn the audio on, as problems in the antennas will usually show up as noise on the audio portion of the signal in addition to the havoc caused in the visual readout.

Tap both sides of the antenna with your broom handle, and go through the rest of the procedures detailed previously—follow through all the way to the display heads. Take special care to check both sides of the antenna, as the VOR's accuracy depends on having both sides of the antenna intact and functioning. One broken side can cause major errors in the readout, especially where the signals tend

Fig. 4-21. Edo-Aire's RT-563-A. (Courtesy Edo-Aire)

Fig. 4-22. A combination nav-com antenna. (Courtesy Edo-Aire)

to bounce off the terrain, as in mountainous areas or when the airplane is banked.

If after looking for the problem in the more likely and owner-repairable components of the VOR, you still haven't found it, you are not yet ready for the repair shop. This is not to say that I recommend unlimbering your soldering iron and attacking the black boxes, but only that again we have come to the point that a little bit of troubleshooting on your part can result in considerable savings in both time and money. Carefully go through the checklist for the VOR, noting any divergencies or differences from previous readings.

As with the engine, your running record will be the best indicator of where the problem is as well as the best indicator of potential problems. A two-degree error in the VOR readout is tolerable, as long as you know about it. When this error suddenly changes, it is a sure sign that something is going awry. You may not know which component is at fault, but your avionics repairman should be able to go right to it rather than spending his expensive time trying to find out what you already know. All you have to be able to do is define the problem accurately and point him in the right direction.

If the problem is intermittent, occurring at a particular time or when the airplane is in a particular configuration, try to pin point it

Fig. 4-23. A com antenna too close to the other accessories. (Courtesy Edo-Aire)

as accurately as possible. Use your imagination as well as the checklist.

Glideslope, Localizer, and Marker Beacon

The landing aids are generally not very troublesome because

Fig. 4-24. A com, probably too close to the rotating beacon. (Courtesy Edo-Aire)

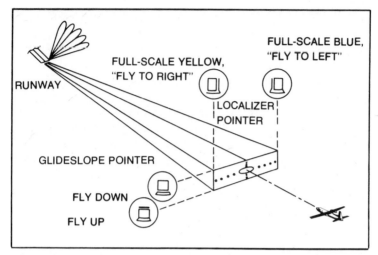

Fig. 4-25. Glideslope information. (Courtesy FAA)

of their simple function and short range, but they make up for this by being almost impossible to troubleshoot without special signal-generating and other bench equipment.

The glideslope is a simple 40 channel receiver that takes the signal from the ground station and converts it into information to drive the glideslope needle (Figs. 4-25, 4-26). It is designed to operate at very short range, and as a result is relatively trouble-free. It is, however, subject to terrain inaccuracies very much like the VOR receiver. Snow and ice near the ground station will cause inaccuracies, as will propeller modulation (which is more likely to disturb the glideslope than the VOR receiver, due to the nature of the glideslope and the airplane's relation to the ground station while landing).

The glideslope is checked out exactly as you would the VOR receiver, with one word to the wise. Due to the nature of the glideslope, if you are checking out the glideslope width by flying the aircraft down the glideslope, be sure you have a competent backup pilot with you. It is all too easy to get so engrossed in what you are doing with the instruments and forget that you are flying the airplane very close to the ground.

The up-down divergence of the glideslope should be about seven-tenths of a degree, or 250 feet for full needle deflection at the outer marker and 35 feet at the middle marker.

The localizer is a part of the VOR receiver and shares its problems, although the short ranges at which it works tend to minimize them, as with the glideslope. Remember that the OBS does not affect the localizer, so when you are on course, you must be in one place. The localizer, a part of the VOR receiver, may work although the VOR itself is not working, and vice versa.

At 2000 feet you should be able to receive the localizer ground station 20 miles out. Full deflection of the needle should be 2.5 degrees or 1500 feet from the centerline at the outer marker. This course width is important, as if it is too wide, the localizer will be inaccurate, while if the course is too narrow, it will be difficult to stay on course. Any sudden change in the course width is a harbinger of problems to come with the localizer.

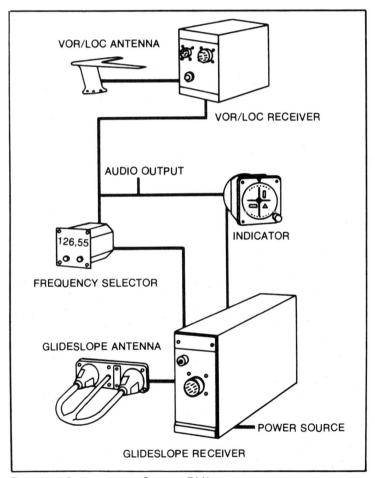

Fig. 4-26. ILS components. (Courtesy FAA)

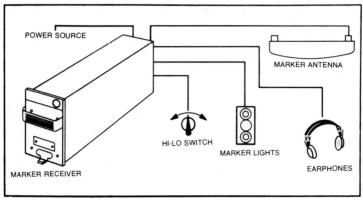

Fig. 4-27. Marker beacon components. (Courtesy FAA)

The marker beacon receiver (Fig. 4-27) has the distinction of being the simplest radio in the airplane. An automobile radio is far more complex than the marker beacon receiver. Being so simple and operating at even shorter range than the other landing aids, it seldom causes any trouble. The lack of trouble is lucky, because it is almost impossible to troubleshoot without a signal generator and other specialized equipment.

The radio receives a very narrow signal sent directly up by the ground station, turns it into an audio tone (through the VHF receiver, which must be turned on) and lights the correct light to indicate outer, middle, or other marker.

Pilots are sometimes surprised when the marker beacon goes off in the middle of a flight, when they are nowhere near an airport. Unless this happens frequently, and in different locations, don't worry about it. It is usually television Channel Five which operates at very high power on nearly the same frequency. When landing, the interference is never a problem, as the FAA has yet to locate an approach over a TV antenna, so the chance of misinterpretation is minimal.

Familiarize yourself with the length of time the marker beacon flashes (and/or buzzes) in your installation and use this as a benchmark. As with everything else on the airplane, a sudden change here may indicate potential failure. If you feel the time the marker beacon indicates is either too long or short, the sensitivity usually can be adjusted. About the only troubleshooting that you can do yourself on this receiver is a visual inspection of the antenna and connectors.

Automatic Direction Finder (ADF)

The old (and not so reliable) ADF, the navigation instrument that made flying as we know it today possible, has in recent years been replaced as the primary navigation instrument for serious flying by the VOR. However, it still has its moments. In fact, serious instrument flying is impossible without it. In addition to its use for approaches, it is a necessary backup for the VORs and for long-range navigation where the VORs are thinly spaced—not to mention its ability to report the ball scores (Figs. 4-28 through 4-30).

The ADF operates in the 190 kHz to 1750 kHz range. This range of frequencies is assigned by the FCC to a multitude of users and purposes, including the AM commercial broadcast band (550-1600 kHz). Among the other uses for these frequencies, are the nondirectional beacons used for aviation navigation and approaches.

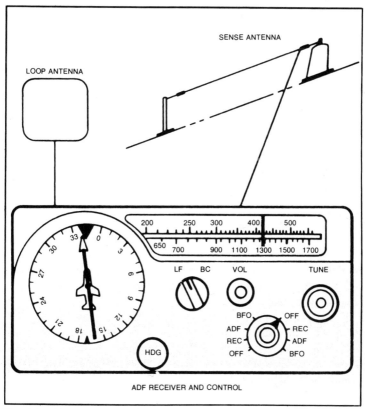

Fig. 4-28. ADF installation. (Courtesy FAA)

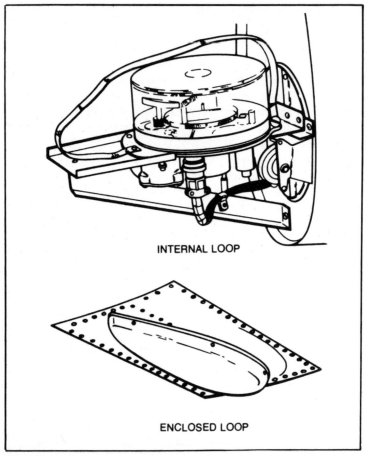

INTERNAL LOOP

ENCLOSED LOOP

Fig. 4-29. ADF loop antenna. (Courtesy FAA)

The fact that the ADF works over a broad band of multipurpose frequencies gives the pilot a generous choice of signals with which to navigate. This was particularly important in the early days of aviation, when navigation beacons were few and far between. Today it allows us to use commercial radio stations as a backup to VOR flying in addition to supplying us with other necessary aviation data.

Today, the most serious ADF navigation is done with the nondirectional beacons set up for aviation navigation. These are frequently at airports, to provide beacons for non-precision approaches. Since these beacons are cheap and adequate (although not as good as precision approaches), they will probably be with us for a long time to come.

The ADF is a simple receiver, again, much like a comm receiver. The signal from the antennas is amplified, processed, and sent to the display head, where a needle always points to the station being received. In the ADF, the most usual complications are the antennas. It requires two. One is a loop which receives the station, and is electronically tuned to the station to give a bearing. The only problem with the loop antenna's bearing is that it contains a 180-degree ambiguity—that is, with only this bearing, you can't tell whether the needle is pointing directly towards or directly away from the station. To eliminate this ambiguity, another antenna is added. This added sense antenna resolves this ambiguity, and causes the needle to always point towards the station being received, thus making the ADF a more useful instrument for navigation.

As with the VOR receiver, there are many problems that appear to be equipment-related but aren't. Due to the low frequencies at which the ADF operates, there tend to be more of these, but they are generally easily anticipated and compensated for.

Noise, of the electrical sort, tends to send the needle spinning in useless circles or pointing in the wrong direction. In an electrical storm the needle will try to point towards the lightning or other electrical disturbance. In rain or snow, precipitation static can build up, and the ADF is particularly subject to atmospherics, due to its

Fig. 4-30. ADF receiver, Edo-Aire's R-556. (Courtesy Edo-Aire)

low frequencies of operation. All these things cause noise, which disrupts the ADF. If you have any doubts about the trustworthiness of the ADF bearing, listen to the audio portion; you can always hear the noise. Static from electrical storms and precipitation can be heard on the speaker, as well as static from faults in the antennas and equipment.

Again, because of the low frequencies involved, the ADF is more subject to terrain effects than the VOR. Fifteen degree errors are not uncommon in mountainous terrain, as the long waves bounce off the mountains to arrive at the receiver with as many shadows as the TV picture in a big city. ADFs are simply not reliable for long distance precise navigation in mountainous terrain, and no number of trips to the avionics repair shop will rectify this fault.

Shorelines tend to bend the signals, giving false readings that can be many degrees off, particularly when approaching a station at an oblique angle to the coastline (Fig. 4-31). This will be most noticeable when actually crossing the shore; the ADF needle will swing to a different heading as the shore is crossed. This bearing, incidentally, will be the accurate one; the previous bearing was bent.

Nighttime navigation with an ADF can be confusing. Long wave signals travel enormous distances, particularly at night. These long-ranging signals skip off the ionosphere, confusing the ADF as they interfere with closer signals on the same frequency, or make the needle point to where the signal bounced from, rather than where it originated. Skip reception is usually not detectable in the audio portion of the ADF—it usually sounds perfectly normal—but interference is. You will be able to hear two or more stations, which will tend to beat as they mix, much as out-of-sync engines on a twin. You have probably heard this phenomenon on the upper frequencies of the broadcast band on the radio in your car or at home.

To reduce skip reception, use the lowest frequencies possible. The ADF beacons are at the lowest end of the frequency spectrum and are more reliable than the broadcast stations on the upper end. Even the broadcast stations on the lower end of the broadcast band are less subject to skip and interference than those in the higher end. From this information, we can come to certain conclusions about navigation by ADF in troublesome times. Avoid navigating by commercial stations. If you must, stay at the lower end of the band. Always keep the range as short as possible; the further you reach for the station, the more likely is it to be inaccurate.

ADFs are also subject to other atmospherics. These can usually be heard on the audio portion, acting much as they do on the VOR receiver. Listening to the audio is your best defense, as is keeping the range as short as possible. Navigating with an ADF when conditions are tricky can be done, but requires knowledge of the set's capabilities and shortcomings.

Troubleshooting the ADF is child's play compared to troubleshooting the VOR receiver. Most of the problems center on noise, most of which is in the antennas and related systems. These parts of the ADF additionally seem to be particularly subject to improper installation and misuse.

The sense antenna must be of the modern insulated sort if the ADF is to be used for instrument flying, and it is most important that the insulation is in perfect condition. One little crack here and the resultant static when it gets wet may make the receiver totally useless. The sense antenna seems to be the most usually bungled antenna at any tiedown area. They look like clothes lines and tend to be treated with the same degree of respect. If yours breaks or needs repairs, and you don't know exactly how it should be done, please don't make the sort of repairs shown in Figs. 4-32 through 4-34; they could be deadly. All these photographs were taken at one small airport!

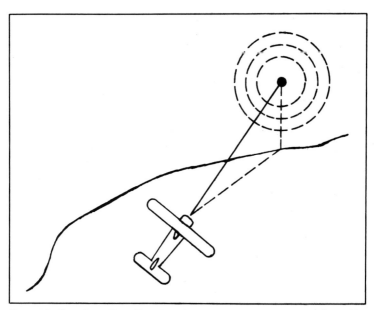

Fig. 4-31. Shoreline effect. The dotted line represents course as indicated by ADF, solid line represents true course.

Fig. 4-32. Faulty installation and repair.

The loop antenna (on modern installations) tends to be encapsulated in plastic. Check it for cracks and signs of water, as it has an old-fashioned vacuum tube-type connector which is particularly subject to water damage. The older style loops should be checked for cracked insulators, loose wiring, etc., but it is far more complicated than it looks. If you aren't sure of what you are doing, have your avionics man look at it.

Tap the antenna, following the wiring and connectors inside as you would for the comm set, and look for problems in the display, while listening to the audio for noise and static. The set should be tuned to a strong nearby station while checking the antenna and associated circuitry.

Since noise is such an important factor, it stands to reason that abnormal noise from aircraft components will cause trouble also. Antenna cable routing is very important to the ADF installation. Look for noise caused by electric flight instruments as well as the more usual electrical accessory motors, ignition, and alternator (generator). The results of static on the ADF are similar to the VOR, but more prevalent and more difficult to eradicate.

Turn the receiver to a strong nearby station and turn every conceivable noise-producing component on and off, remembering that electrically-operated gyro instruments can usually only be killed by using the master switch. Run the engine, switch magnetos,

Fig. 4-33. Vacuum tube-type connectors tend to be subject to water infiltration. They should be cleaned and packed with Vaseline to minimize this problem. Antenna side. (Courtesy Edo-Aire)

etc. Finding an offending component often discloses a problem with it rather than with the ADF, although failed shielding and other problems can allow noise in the system that would otherwise not affect it.

Fig. 4-34. The fuselage side. (Courtesy Edo-Aire)

Fig. 4-35. A transponder, Edo-Aire's RT-777, TSO. (Courtesy Edo-Aire)

Transponder

The fury surrounding the transponder is dying down. We are all getting used to having the expensive little tattletale with us, and even beginning to see it as a useful flight tool. An airplane without a transponder is (as of this writing) almost useless in busy areas, a situation unlikely to change as years pass (Figs. 4-35, 4-36).

Most air traffic controllers are polite and exercise tact when noticing our altitude divergencies. Rather than bluntly telling us when we are out of our assigned altitude, they will hint with barometric information that allows us to correct and save face at the same time. In these days of increased crowding of the airways, it can be a comfort to know that we are well seen by the controllers.

The transponder seems quite a simple gadget, but in fact it is quite complex. Along with the DME, it is the most sophisticated piece of equipment likely to be in an average general aviation airplane. It is both a receiver and transmitter, operating through a common antenna on a fixed frequency, 1090 MHz transmitter, 1030 MHz receiver. Remember this, for when you are changing your identifier code, you are changing the modulation, not the frequency.

The transponder operates at very high voltages. It transmits (only when interrogated by a ground station) at between 125 and 500 watts (depending on the transponder) compared to 20 watts for the usual VHF transmitter. It can transmit at such high power because it transmits on a pulse basis for brief periods—.45 millionths of a second. It is very effective in bad weather and poor radio conditions because of the extremely high frequency and the high power used for transmission.

Because of the transponder's nature, most of the problems that you will have with it will be reported to you by a controller. This puts you at an immediate disadvantage in troubleshooting. You are at once removed from the problem before you even get started. Try to find out from the controller exactly what is and is not happening. Don't be afraid to ask him specific questions if he is not too busy. At the same time, carefully note any factors in the operation of the airplane that may be affecting your transmission, as well as checking the set itself. It is important to have a complete, explicit report for the avionics repairman.

As usual, the best (and cheapest) place to start is the antenna and associated circuitry. The transponder antenna is short (two inches) and looks like a sawed-off automobile antenna, but can also be plastic encased, and is mounted on the belly of the airplane. This, of course, is just where it gets most benefit from whatever oil leaks you may have in the engine or hydraulic system. Excess zeal in the cleaning of the airplane frequently results in a bent, broken, or otherwise damaged antenna. If it is bent or covered with grime, it is not working properly.

Fig. 4-36. A transponder to fit an instrument panel hole and save radio rack space for more interesting stuff, the Edo-Aire RT-887, also TSO'd. (Edo-Aire)

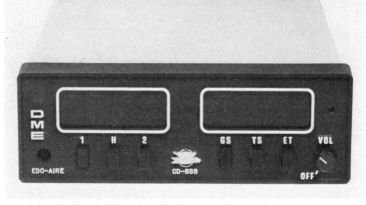

Fig. 4-37. Edo-Aire's CD-880 DME. (Courtesy Edo-Aire)

The transponder does not respond very well to antenna tapping with your broomstick, wire pulling and such, due to the nature of the beast, but a careful visual and physical inspection of the antenna and wiring will frequently reveal faults. Once this is done, there is little more the pilot can do other than make sure the mechanical parts are functioning properly. The antenna wire, length, type and condition are vitally important in the transponder installation, as the high wattage used to transmit can wreak havoc with the other avionics if there are any faults or defects.

The transponder is probably the most difficult single piece of avionics equipment to install properly, so make sure it is functioning properly when the installation is new. Improper antenna cable or incorrect length can cause a voltage shift in the transmitter that will render the transponder almost useless. With 4096 codes to choose from, it is likely to be a long time before you find out that one is not working. As the set grows older, losing power, installation faults may appear that weren't obvious when the set was new. The installer's solutions to the problems of antenna location and cable routing which were adequate when the installation was new can be inadequate as the set's power declines in old age. Of course, this weakening with age affects other avionics as well, but it is far more pronounced in the transponder because of its high output wattage, which makes it very critical in this area.

Remember that the transponder signal may be shadowed, either by banking the aircraft through the interrogating signal, or by a badly placed antenna that is blocked by an opening landing gear door, rotating beacon, or other antennas. Installing a transponder is the height of the avionics installer's art, and requires the highest degree of expertise.

Although not uniquely, transponders are particularly subject to high altitude arcing. High altitude arcing is caused by the reduction in resistance that occurs at low atmospheric pressures, which tends to allow electric current to flow across contacts and weaknesses in insulation at high altitudes. Because the phenomenon is present at high altitudes only, the problem is never evident on the test bench. The transponder is particularly subject to this phenomenon because of the high voltages present in the set. If you have a problem at high altitudes with a set, particularly if you have just had some work done on the set, suspect this as a possibility. Insulating coatings that have to be removed from components within the set to repair it must be very carefully replaced to avoid this problem.

Not all high altitude problems, however, are due to high altitude arcing. There is always the possibility that the colder temperature at higher altitudes is affecting the performance of the set, particularly if the set is remote-mounted in an unheated part of the airplane.

Distance Measuring Equipment (DME)

The DME, which seemed like such a luxury a few years ago, has quietly slipped into the area of a necessity. Part of the reason for this is the reduction in price, part a fuller use of the Air Traffic Control System by even the non-instrument pilot, and part human nature. The DME has more entertainment value than any other item on the airplane!

The DME is a transmitter and receiver that works a little like a reverse transponder (Fig. 4-37). It sends a signal interrogating a ground station, and the ground station sends a signal back. The total time between the moment the signal is sent from the DME to the time the DME receives the signal from the ground station is measured and converted to useful distance and time information. A computer in the DME then figures the speed. The airborne transmitter is transmitting all the time, while the ground station equipment transmits only when interrogated by the ground equipment. The different ground stations have different frequencies between 987 and 1213 MHz.

Like all other avionics, knowing just what the DME will and will not do is important in avoiding unnecessary service calls. As we go through the strengths and weaknesses of the DME, you will see that most of them are reflected primarily in the speed function, which should tell you something about the accuracy and reliability of that function.

The DME measures the distance between it and the ground station at fixed intervals. From the closing speed information, it computes the groundspeed of the airplane. This calculation grossly accentuates any errors in input. There are three major sources for calculation errors other than equipment malfunctions.

First, by its nature, the distance calculation is really only an approximation—close enough for distance measurements, but too approximate under certain conditions for really accurate groundspeed calculations. This is because the DME distance is not the actual ground distance of the airplane from the DME, but the real distance of the airplane from the DME. The DME distance of the aircraft from a given ground station is the hypotenuse of a right triangle whose other legs are the ground distance and the altitude of the aircraft. As long as the distance of the aircraft from the beacon is large, the hypotenuse (direct line distance) and the ground distance are, for all practical purposes, the same, but as the distance becomes smaller, the error increases to the point that directly over the beacon, the error may be said to be total. That is to say, the DME is measuring the *altitude* of the aircraft rather than the ground distance from the station at that point. As long as we are careful to specify ("Five miles DME from Holston Mountain VOR"), controllers, and we hope pilots, understand, but the electronic mind of the computer doesn't, and it only sees the speed of the airplane dropping as we approach the station, suddenly accelerating as the airplane passes over.

Second, the ground station has a period (about five seconds) when it transmits its identification. During this period, it is not transmitting replies to interrogations from airborn DMEs. The DME has a memory to carry it over these identification periods. It simply remembers the last readout and displays it until a new one is received, creating for these brief periods an infinitesimal error of no consequence to the distance reading. Not true of the groundspeed calculation. Since it is not receiving its time signals, it thinks the airplane has stopped dead in the air and starts unwinding to indicate this condition. Since the readout is very heavily damped, it never really gets to show you this misinformation, but the average DME will lose 10-15 knots before the signals resume and start it winding upwards again. DMEs react differently to this phenomenon, but they all react.

And third, the groundspeed calculation is only accurate if you are approaching the station dead on. If you are just a little bit off course, especially as you get close to the beacon, the geometry goes against you again (just as it did in school), and the calculation becomes grossly inaccurate.

All of this is not to say that the groundspeed indication is useless, only that it should be understood and used properly. It is at its most accurate when the aircraft is at a reasonable distance from the ground station, exactly on course, and the reading has stabilized from initial warmup or a ground station identification period. The groundspeed display is heavily damped to avoid constant changes in the reading, and can take from one to three minutes to stabilize.

From the previous paragraphs, it is obvious that a minor malfunction in the DME, one which will not appreciably affect the distance reading, can cause inaccurate or erratic groundspeed readouts. Thus, very often, readings of this nature can be harbingers of troubles to come in the DME. Therefore, a good description of groundspeed errors may be helpful in diagnosing a later failure in the distance display.

The antenna is, as usual, the first place to look for trouble. Sometimes the antenna is the same sort as the transponder, in the same location, subject to the same problems; sometimes it is a fin-type plastic antenna. Antenna leads and cable routing are important. Noise picked up by improper or improperly routed antenna leads can cause erratic groundspeed indications. Improperly shielded or routed antenna leads can also generate troubles in other avionics. The DME is a relatively simple installation, but the antenna should not be shielded by gear doors or other installations, as it is a low-power installation and needs all the help it can get.

The DME is also subject to voltage fluctuations. Low voltage makes a very small change in the distance readings, but comes back to haunt us in the speed readout. Low voltage problems can be checked for by shutting off the alternator/generator, while looking at the speed display for a change in readout, allowing for the heavy damping in the display. Low power will also cause a reduction in the already marginal DME transmitter output, which may appear to be an equipment failure.

As with the other avionics that are difficult to diagnose without special service equipment, the secret to repairing your DME as cheaply and quickly as possible will be describing the symptoms as clearly as possible to your avionics repairman, after first checking out the antennas, connectors, and leads yourself.

Emergency Location Transmitter (ELT)

The ELT is a simple radio, broadcasting on one frequency with a self-contained battery. There is not too much to be said about the ELT. The problems here are in the batteries, and nobody seems to have the definitive solution to the problems. Look at the ELT occasionally to make sure that the batteries are not in any way destroying your airplane. Make sure the antenna is located where it will not blanket or cover any useful antenna. Listening for it on 121.5 MHz is a part of the avionics checklist earlier in this chapter. This should be done to make sure it hasn't decided to turn itself on. (It probably won't if you have an accident and really need it. But that's all right, nobody seems to find it even if it does.)

Chapter 5

Do-It-Yourself

Doing it yourself is the most obvious approach to saving money in the operation of your airplane. Unfortunately, it requires more common sense or elbow grease. If you are the sort of person who calls the electrician to change a light bulb, or the garage to change a flat on your automobile, you probably aren't capable of doing much on your airplane, either.

A lifetime of doing minor repairs on your automobile, house, and lawnmower is good preparation for doing the necessary work on your airplane. If you have no inclination for this sort of thing, you should probably stay away from your airplane. On any warm Saturday morning, the airports are full of pilots pulling maintenance on their airplanes who wouldn't think of doing the same on their houses, automobiles or other mechanical equipment. Why do they work on their airplanes? The obvious reason is to save money.

One of the many reasons that aviation maintenance costs are high is that there is no margin for error. A plumber's work fails and you get wet—an A&P's work fails, and there is the possibility of real danger to your life. The A&P's training is long and expensive; in addition, he and his license are under constant pressure from the FAA to maintain extraordinarily high standards. His work is done, all the time, to a necessarily high standard.

You should remember this when you are deciding just how much you want to do and how far you want to go in maintaining your own aircraft. How good are you? The feeling that the FAA wouldn't allow you to do it if it could get you into trouble is simply not so. The FAA gives you more than enough rope to hang yourself if what you do is not done thoroughly and correctly. In an airplane there is no room for loose connections, improperly torqued nuts, or forgotten steps. Slipshod work may extract a heavy price.

Part 43, Appendix A, (c), (1) - (25) of the Federal Aviation Regulations specifies exactly what the FAA considers preventive maintenance and therefore within the scope of the owner/pilot to perform on his own airplane. This list applies only to airplanes not flown for hire. That the FAA says that you may do the following things does not mean that you have either the desire or the ability to do them. Some of the work is of the frustrating, thumb-bending, knuckle-busting sort, some takes an incredible amount of time without highly specialized equipment, and some is well beyond your abilities, unless you are very handy indeed. You must be careful not to exceed your capabilities, and you should never be too proud to ask for help if you are not sure of a procedure.

Doing your own preventive maintenance is not all free. You will have to build up a toolbox, and today that requires a considerable expenditure. For some jobs, you will have to buy a special tool. Weigh this purchase in the decision of whether you want to do the job yourself. Sometimes it is cheaper to have it done, rather than buy a specialized tool.

If you do need a special tool, buy it. Do not be a tool borrower, particularly from the A&P who bought the tool and would otherwise be getting your work to help him pay for it. Tool borrowers are anathema at any airport or garage, and they are (and indeed deserve to be), grouped with sufferers from BO, bad breath, and ring-around-the-collar as social outcasts.

The following titles (in italics) are the exact wording and numbering of the paragraphs of Part 43. Each paragraph is headed by a list of special tools necessary to do the job. This list is followed by a brief description of what should and should not, may and may not, be done. My opinion about the job, whether it represents a good investment of time and bashed knuckles, is sometimes included.

This chapter is not a manual and should not be used as one. Its purpose is to help you decide whether or not you care to tackle the operation described. You should have a copy of the maintenance manual for your airplane before you actually start the work.

LANDING GEAR

(1) Removal, installation and repair of landing gear tires.

Tools: Suitable jack, torque wrench, aircraft tire bead breaker.

Note that the wording is "landing gear tires," not "landing gear and tires". Other things you may do to your landing gear are covered in other paragraphs, and very specifically.

This operation is considerably more complicated and critical than changing the tires on your car. It requires demounting the tire from the wheel, an operation most of us are content to let a garage do on our automobiles. In the case of an airplane wheel, the tire can be very difficult to remove, and the very expensive wheel can be easily damaged in trying to remove the tire (Fig. 5-1). Demounting the tire is most easily done with special bead-breaking equipment (similar to that used in a garage), although it certainly can be done on the pavement next to the airplane.

To inspect the wheel and tire assembly, the airplane must first be jacked up (Figs. 5-2, 5-3). Be careful to jack the airplane at the proper spot—a mistake here can be very expensive. If you are not sure exactly where the jack should be placed, ask. Once the wheel assembly is jacked up, remove the wheel pant (if applicable) and inspect the tire for tread wear, abrasions, and cuts, as well as ozone deterioration of the sidewall. (This last applies if the airplane is infrequently and/or very carefully flown. It is possible to have a tire wear out of old age.)

If the tire requires replacement or repair of the wheel, remove the wheel nut, bearings, and brakes, if necessary. Let all the air out of the tire by removing the valve core, and break the bead from the rim, being careful not to scratch, gouge, or otherwise damage the

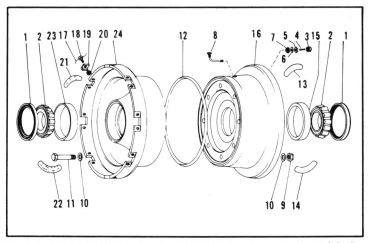

Fig. 5-1. A typical light aircraft split wheel. It is almost as complex as it looks.

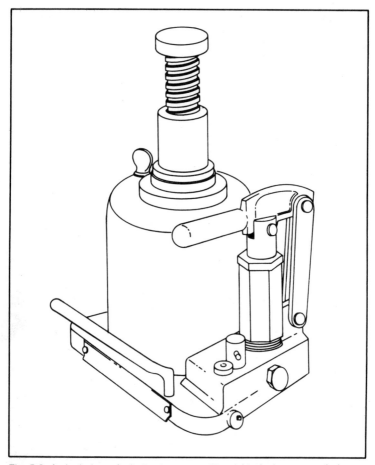

Fig. 5-2. A single base jack, the type you will need to jack up your airplane.

wheel in the process. (An insignificant-looking gouge could render the wheel useless.) Once the beads of the tire are broken from the wheel, disassemble it. Tubes and tires may be repaired in the usual manner (do not use automotive-type plugs in tubeless tires), but if the tire is at all doubtful, have an A&P pass on it. Tires can be recapped, and this can work out well, particularly if the airplane is hard on treads, either by its use or design.

If the tire is tubeless, there may be an "O" ring between the wheel halves to retain the air. Make sure this is sound and properly located when reassembling the wheel. If there is a tube, reassemble the tire and rim carefully, dusting the tube with talc, and making sure it is not pinched anywhere. Line up the mark on the tire (the light point) with the mark on the tube (the heavy point), and assemble the two halves of the wheel, taking special care not to pinch the tube of a standard tire, or ensuring that the "O" ring is in place in the case of a tubeless tire. The halves of the wheel should then be carefully torqued to the recommended value. Inflate the tire to the recommended pressure, let the air out by removing the valve core, and reinflate the tire. In the case of a tubeless tire, if there is a mark on the tire, it should be located next to the valve. It is good practice to renew the valve when the tire is replaced. If there is a balance problem with the completed assembly, try rotating the tire 180 degrees on the wheel—this often helps.

The wheel assembly should now be carefully replaced on the airplane, with reference to item four as to the wheel bearings. While you are doing this, remember an error here would most probably not be fatal, but the wheel could come off or seize, or the tire could go flat, and any of these could be very expensive.

As a matter of fact, I recommend having someone else do this job. It is vastly simplified by having the proper tools and equipment, and the price is usually quite reasonable. The mechanic can accomplish in fifteen minutes what it may take you all afternoon and three bashed thumbs to do.

After tire repair or replacement it is best to sit overnight before flying. Then, if a leak is present, the tire will go flat on the ground instead of in the air.

(2) Replacing elastic shock absorber cords on landing gear.

Tools: A suitable jack, chain hoist, and special fixture for stretching the cord.

This can be a very nasty job, particularly if it has to be done on the airplane. If your airplane has a demountable shock strut assem-

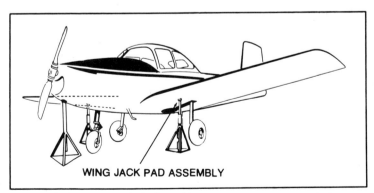

WING JACK PAD ASSEMBLY

Fig. 5-3. Jacking up the whole thing. More equipment, more complexity.

bly, it is relatively simple. Jack up the airplane, remove the demountable assembly, and let your A&P replace the cord on the assembly. If the airplane is designed so that the work must be done on the airplane, let your A&P have the whole job. Stretching the cord is a job for a chain hoist and/or fixture.

Incidentally, the shock cord is color-coded for the date of manufacture. This is important to know, as the cord deteriorates, even on the shelf, and you don't want to replace old cord with old cord.

(3) Servicing landing gear struts by adding oil, air or both.

Tools: Clear plastic tubing to fit over valve, container of recommended hydraulic fluid.

The hydraulic strut, which fulfills the same function as the shock cord (but with more sophistication), contains hydraulic fluid like an automotive shock absorber, and is pressurized with air over the oil. It is important that the strut be filled to the proper level with oil, and the air be at the correct pressure.

The only real difficulty in doing the job yourself is obtaining a high-pressure air source to fill the strut. Borrowing air may or may not be construed as tool borrowing; let your conscience be your guide.

A shock strut is made up essentially of two telescoping cylinders or tubes, with externally closed ends. These comprise the cylinder and piston, forming an upper and lower chamber for movement of the fluid. The lower chamber is always filled with fluid, while the upper chamber contains compressed air. An orifice between the chambers provides a passage for the fluid into the upper chamber during compression of the strut, and a return during extension of the strut.

Most shock struts employ a metering pin to control the rate of fluid flow into the upper chamber. On some types of shock struts, a metering tube is used instead of a variable shaped pin, but strut operation is the same.

Some shock struts are equipped with a damping or snubbing device consisting of a recoil valve on the piston to reduce the rebound during the extension stroke.

A fitting consisting of a fluid filler inlet and air valve assembly is located near the upper end of each shock strut to provide a means of filling the strut with hydraulic fluid and inflating it with air.

A packing gland designed to seal the sliding joint between the upper and lower telescoping cylinders is installed in the open end of the outer cylinder. A packing gland wiper ring is also installed in a groove in the lower bearing or gland nut on most shock struts to keep the sliding surface of the piston free from dirt, ice, snow, etc.

The majority of shock struts are equipped with torque arms attached to the cylinder and piston to maintain correct alignment of the wheel. Those shock struts without torque arms have splined piston heads and cylinders.

Nose gear shock struts are provided with an upper locating cam attached to the upper cylinder and a mating lower locating cam attached to the piston. These cams line up the wheel and axle assembly in the straight-ahead position when the shock strut is fully extended. This prevents the nosewheel from being cocked to one side when the nose gear is retracted, thus guarding against possible damage to the aircraft structure. These cams also keep the nosewheel in a straight-ahead position prior to landing. Some nosewheel shock struts are fitted with shimmy dampers.

All shock struts are provided with an instruction plate or decal which gives directions for filling the strut with fluid and for inflating the strut with air. But many of these disappear with time, and in any case it's always best to refer to your airplane's maintenance manual.

The strut must be fully deflated to add oil, and you will jack the aircraft and allow the strut to be fully extended before deflating it. If you deflate the strut while it is compressed, a geyser of oil will blow out through the valve. Also, extended or not, keep your face away from the strut valve while deflating it. That's high-pressure air, and it can be dangerous to your eyes.

It should be noted that, although the strut valve core may greatly resemble the valve cores in your tires, they are not the same. The valve core in the shock strut has neoprene seals, which are impervious to deterioration from oil.

We don't recommend that you remove the core from the strut valve to deflate, but instead control the deflation by releasing the air pressure slowly.

With the shock strut fully extended and deflated, attach a length of plastic hose (the kind your doctor uses on intravenous bottles works nicely) to the valve and immerse the opposite end of the hose in your can of hydraulic strut oil. Then, by alternately compressing and extending the strut by hand, simply pump in the fluid to the proper level (Fig. 5-4). If the strut won't pump fluid, it needs new seals.

This method will work equally well in adding fluid to the main wheel shock struts. Keep in mind, however, that while the normal air pressure in most lightplane nosewheel struts is less then 100

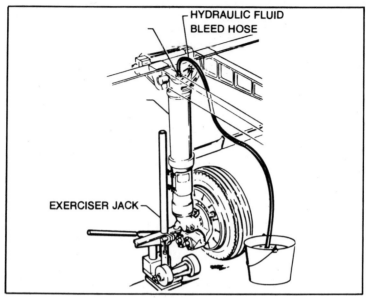

Fig. 5-4. Bleeding the oleo with an exerciser jack.

lbs/sq in., the main wheel struts may carry up to 500 lbs/sq in. pressure, which is why a strut pump or other source of clean, high-pressure air is needed to inflate main gear struts when they are compressed with the weight of the airplane.

However, since there is relatively little air in a shock strut, the 100 lbs or so pressure that your portable air tank will handle may be sufficient if you inflate the main gear strut while the wheel is jacked and the strut fully extended.

While the airplane is jacked, check the bolts and connections on the strut's scissors arms for looseness. This is the source of a lot of vibration.

(4) Servicing landing gear wheel bearings, such as cleaning and greasing.

This operation is generally necessary when removing the wheel to change a tire. The axle construction is similar to the front axle of your automobile, and is handled in the same way. The parts are cleaned, inspected for wear, brinelling, pitting, or any other defects. This inspection is one of the main reasons for repacking these bearings, so do it carefully. Replace worn seals or doubtful bearings, and repack the bearings with an approved lubricant, being careful to maintain a high degree of cleanliness. (We are, after all, packing the bearings with grease, not sand.)

The bearing is then reassembled on the axle, the wheel replaced, and the wheel nut tightened to the specification in the aircraft's manual. The tightening of the wheel nut determines the load on the bearing, so it is very important. Be careful replacing the cotter pin. Refer to the following section.

SAFETY WIRING

(5) Replacing defective safety wiring or cotter keys.

There are two important things to remember about cotter keys (or pins) that are frequently forgotten or overlooked. The first is that they are never reused, the second is that there is a correct size cotter pin for each hole; that is, the diameter of the pin should be a reasonably snug fit in the hole provided for it in the bolt. The ends are trimmed to fit, neatly, so as to leave no sharp ends to cut unwary fingers later. The ends are twisted, one over the top of the bolt, the other down the side of the nut, and snugged up against the fastener (Fig. 5-5). The ends may also be spread around the circumference of the nut, but this method is frowned upon by those who have pride in their work.

Safety wiring is hard to get really neat without a special tool to twist it, but an adequate job can be done with a pair of pliers. A larger job can be speeded up by using a hand drill, but the ultimate job is rarely achieved without a special tool. This tool is not cheap (it's hardly worth buying for one or two jobs), but the professional mechanic takes great pride in the appearance of his safety wiring. It is the external sign of work well done on the inside, and if you want to match him, you will almost have to have one. If you plan to check your own oil screen, or any other job that requires safety wiring on a regular basis, it is easy to justify the purchase of one of these tools.

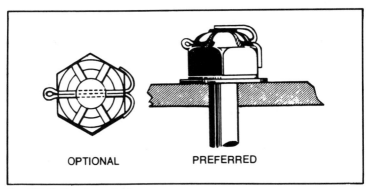

OPTIONAL   PREFERRED

Fig. 5-5. Cotter pin methods.

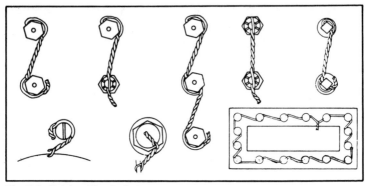

Fig. 5-6. Safety wiring techniques.

Safety wire is a special sort of wire, most commonly stainless, sometimes brass, and infrequently these days, galvanized. Wiring should be done as illustrated, (Figs. 5-6, 5-7), being careful to exert a strong pull on the wire in the direction of tightening. The twist on the wire should be as uniform as possible—neatness counts, and the end at the last bolt should be cut off, leaving at least three turns, and folded neatly against the bolt head. The twisted wire should fill ¾ of the hole drilled in the bolt head.

You are also allowed to replace the safety wiring on turnbuckles, but you are not allowed to adjust them (Fig. 5-8). Probably, if the wiring is incorrect, so is the adjustment, in which case the A&P should check it and complete the safety wiring. If the case arises in which you have occasion to redo the safety wiring, follow the illustration carefully, being particular with the four-turn wrap and in using the right size wire.

LUBRICATION

(6) Lubrication not requiring disassembly other than the removal of non-structural items such as cover plates, cowlings and fairings.

Tools: Grease gun, if required; correct lubricants; oil cans and such.

These FARs were originally written in the early dawn of aviation, or at least the early dawn of the regulation of aviation, and some of these paragraphs show it more than others. This one can cause trouble, as it seems to imply that all airplanes require frequent lubrication. This in general is not true; airplanes have not required this sort of lubrication since the last OX-5 broke a rocker arm because it wasn't properly greased prior to takeoff. (They broke even when they *were* properly greased!)

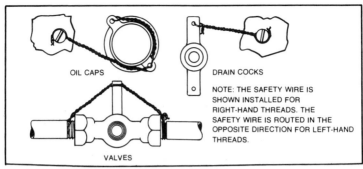

Fig. 5-7. Safety wiring oil caps, drain rocks, and valves.

If your airplane has any spots that need attention between annuals, check your maintenance manual and A&P. Too much grease or oil can cause real trouble. Be sure you know what you are doing.

PATCHING

(7) Making simple fabric patches not requiring rib stitching or the removal of structural parts or control systems.

Tools: Pinking shears and other stuff you can borrow from the family sewing kit, if you don't get caught.

Patching fabric is a pleasant job. It is expensive to have done, and it returns a great deal of satisfaction for the effort expended. The first thing to do is determine what the finish of your airplane is. (Many are painted with enamel over the dope.) If the airplane is

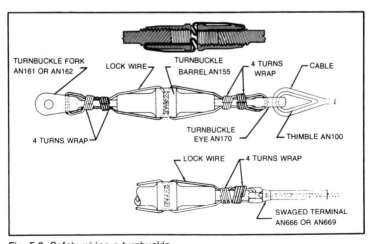

Fig. 5-8. Safety wiring a turnbuckle.

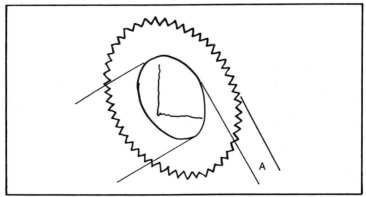

Fig. 5-9. Repairing a small tear. *A* must be at least two inches. Be careful to remove all damage.

finished in enamel, this must be removed by light sanding, as the repair procedure requires softening the dope to hold the patch. Cut out all the damaged area, leaving a round or oval area. With the pinking shears, cut a patch large enough to cover the hole by at least two inches all around. The patch should be made of the same material as the airplane cover (Figs. 5-9, 5-10).

Paint the area around the hole for two inches, plus a bit, with butyrate dope. This will soften the dope on the cover. When it has, scrape the dope away down to the bottom layers. Put another coat of dope on the area and lay the patch in place. Cover with dope in the usual fashion (clear, silver, clear and color) and the patch, if carefully done, will be almost unnoticeable.

HYDRAULIC FLUID

(8) Replenishing hydraulic fluid in the hydraulic reservoir.

This is usually a straightforward operation. Make sure you use the recommended hydraulic fluid and don't overfill, paying particular attention to the airplane's maintenance manuals instructions. If you fill your own reservoir, you will notice if there are sudden changes in fluid needs. They are like most other sudden changes in your airplane—they signify a problem developing. If you suddenly need more than a nominal amount of fluid to fill the reservoir, check for leaks, and check with your A&P before operating the aircraft.

AIRFRAME AND INTERIOR

(9) Refinishing the decorative coating of fuselage, wings, tail group surfaces (excluding balanced control surfaces), fairings cowl-

ing, landing gear, cabin, or cockpit interior, when removal or disassembly of any primary structure or operating system is not required.

Painting an airplane is a specialized job. It has very little in common with automobile finishing, house painting, or any other associated decorative art that you might have acquired over the years. This regulation gives you license to paint almost the whole thing. Unless you really know what you are getting in for, *resist the temptation.* To do a good job, a large interior space, specialized paint removers, spray equipment, and other equipment are necessary. By the square foot, for equal quality, repainting an airplane is considerably cheaper than painting an automobile. Find a good shop and have them do it—*right.*

A poorly-done home paint job can do more to ruin the value of an otherwise good airplane than any other single thing, while maintaining an existing paint job does more to preserve the value than anything else. Your time is better spent in the latter pursuit. However, since 80% of the cost of re-painting is labor, some owners tackle the job themselves. See TAB Book No. 2291, *Refinishing Metal Aircraft,* for detailed techniques.

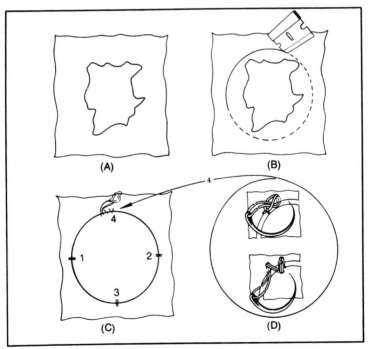

Fig. 5-10. An alternate method of fabric repair.

(10) Applying preservatives or protective material to components where no disassembly of any primary structure or operating system is involved, and where such coating is not contrary to good practice.

This applies primarily to seaplanes and, to a certain extent, airplanes that are kept in seashore areas that may need preservatives applied to exterior bolts, cables or pulleys. If you have a seaplane, consult the manual for preservation recommendations. For a floatplane, consult the float instructions. If you are having corrosion difficulties with an airplane that is based in a salt water or other corrosive atmosphere, consult your A&P before deciding what steps to take to preserve it, or reference the before-mentioned TAB book which discusses corrosion and rust problems in detail.

(11) Repairing upholstery and decorative furnishings of the cabin or cockpit interior when the repairing does not require disassembly of any primary structure or operating system, or interfere with an operating system or affect primary structure of the aircraft.

This allows unbridled license so long as common sense is applied. You can have your upholstery in early Victorian plush, or military spartan, or anywhere in between, so long as you are careful to avoid fouling the controls or otherwise compromising the aircraft's flight performance.

The seats and their attachments may not be changed or modified, as they are considered a part of the primary structure of the airplane. When you are designing your dream interior, remember the seat belts, controls, seat adjustments, and door locks come first. They must be free to function properly; the interior should be modified to fit *them* rather than the other way around. Materials chosen should be incapable of sustaining combustion, which may limit your choices somewhat.

The instrument panel is immutable under this FAR, nor does it give you license to install a teen-dream stereo sound system with your reupholstery job. A stereo or tape deck must be installed by an avionics shop or A&P.

When designing your interior, avoid anything too personal if you care about the resale value of the airplane. Original equipment is usually rather tasteless and plain, but it has the widespread acceptance that makes an airplane easy to sell, which is probably why it was designed that way to begin with.

This paragraph, incidentally, also allows you to make repairs to your existing upholstery, this topic is covered in Chapter eight.

(12) Making small repairs to fairing, non-structural cover plates, cowlings, and small patches and re-enforcements not changing the contour so as to interfere with proper airflow.

This is pleasant spring Saturday afternoon stuff, on a par with patching holes in your fabric-covered airplane. It is the sort of work that requires a large investment in hours, but repays the effort in both money saved and satisfaction from a job well done. Properly done, metal patches look very good and fiberglass patches are unnoticeable.

Cracks have a predilection to developing on engine cowlings and fairings where stress and vibration are the greatest. If you see a crack developing in a non-structural area, it should be stop-drilled on either end with a ⅛-inch drill. This should be done immediately to stop the crack from spreading, even if the crack isn't big enough to patch. If it is big enough to patch, drill it as soon as possible to stop expansion. You can patch it later, when you have the time. A piece of duct tape, if necessary, will keep the crack from leaking. If you are not sure if a piece of metal is structural or non-structural, check with your A&P.

Neat patching, like safety wiring, is an area of pride with a good mechanic. If you want to compete at his level, you will have to be very neat and careful indeed. The legal requirements for a patch are that the rivet centers must be at least two rivet diameters from the crack, as well as from the edge of the patch, and that the patch should be made of aluminum of the same thickness as the metal being repaired. The hardest thing to do in patching is drilling the rivet holes straight and keeping them in a straight line. Extra effort here, and an effort to line up the rivets in the patch with rivet seams already on the airplane will make the patch look professional, or better (Fig. 5-11).

Dents, cracks, tears, etc., in metal aircraft structures are never to be filled in with automobile body putty or filler. *Never.*

Although the techniques used for aluminum airplane repairs are unique to airplanes, fiberglass repairs to non-structural fiberglass surfaces have much in common with fiberglass repairs on boats and automobiles. The cowling or fairing should be removed from the airplane, as the repair is generally reinforced from the back. A close-weave glass cloth should be cut about ½ inch oversize to cover the damage. Three more patches are then cut, each ½ inch larger than the previous one. The area to be repaired is cleaned, sanded and prepared for the resin. The resin is applied and the patches are laid on it. Additional resin is applied between each layer; be careful to express any air from the patch. When this internal, structural reinforcement is cured, the outer side is cleaned and prepared for cosmetic repair. Clean up the area, sand through all

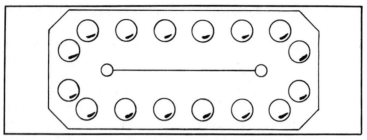

Fig. 5-11. A metal patch, original hole stop drilled, all dimensions correct.

paint and coatings, and apply resin or filler to the area, finishing as you would an automobile.

The actual fiberglass repair material for the patch (cloth and resin) is available, along with detailed instructions for its use, from a boating supply store. Body filler or putty is available from automobile supply stores. Some frown on body filler, even for filling the small cracks left after repairing fiberglass. It is not necessary, as resin may be mixed with chopped fiberglass strands, but the body filler works more easily, and bonds perfectly to the fiberglass.

(13) Replacing side windows where work does not interfere with structure or any operating system such as controls, electrical equipment, etc.

In Chapter 8 care of clear acrylics is discussed. Before deciding to change a window, read that section. There is quite a bit that can be done to freshen up old acrylic; you may be able to save yourself some time and money. There is no cure for crazing, breakage, or discoloring, and under these conditions the acrylic must be replaced.

You can buy acrylic plastic from your local hardware dealer and cut it to size, but I have always found it better to buy blanks from aircraft suppliers. Most of the work is done—there is only a slight danger of blowing the job. Acrylics tend to be brittle; sawing is sometimes successful, sometimes not. The same applies to drilling, but it can fool you—you think you have been successful, and then it cracks later. Usually, the piece you are working on breaks on the last operation, thus wasting not only the acrylic, but also your time.

The installation procedure is very simple. Remove the old, replace it with the new, using the recommended sealer. The window you get from your supplier will not have holes in it; that is your responsibility. It is possible to drill these holes with an ordinary drill, or better yet a special drill made for the job, but the best method by far is to use a hot nail about the size of the fastener. Even if you have an initial success in drilling the holes, they are likely to crack later, long after the window has been installed.

The hole you melt in the window with the hot nail should be generously larger than the fastener, to allow for the thermal stress that may crack the window later. The heating seems to leave the window around the hole less subject to crazing and cracking than drilling the hole through, in addition to the fact that there is no danger of cracking the window while making the hole.

Use the window you have removed to practice and refine your hole-making technique. It is not the sort of job that requires high skill levels. I have found a propane torch to be the ideal heating medium. The nail does not have to be terribly hot, and is best held in your trusty vise-grips. Leave the protective paper or plastic on your window as long as possible; ideally, until after the window is installed. You really don't want any scratches or gouges on your brand-new window. Windows must be fitted slightly undersize to allow for heat expansion.

(14) Replacing safety belts.

Not a difficult job, but a word to the wise. Even though automotive safety belts meet much higher standards than aircraft belts, look better, and cost 1/10th as much, they are *not* suitable replacements for aircraft belts under any circumstances. The only ones approved for aviation use are the ones with the right parts numbers—even if the belt is identical to the one from your local GM dealer.

Also, there are no modifications of the mountings or mounts permissible. No "improvements," please.

(15) Replacing seats or seat parts with replacement parts approved for the aircraft, not involving disassembly of any primary or operating system.

This item does not authorize you to repair the seat once it has been removed, nor does it authorize you to modify the seat. Item 11 allows you to reupholster; this paragraph allows you to get the seats in and out of the airplane so you can do it. If any repairs, such as welding or riveting of the seats is necessary, it must be done by an A&P.

When replacing the seat, be especially careful in fitting it to the rails, making sure that the locks function correctly, so the seat will not slip when taking off. This can be very dangerous. When a seat has been reupholstered, check to make sure that the new upholstery has not fouled the operating mechanism of the seat.

(16) Troubleshooting and repairing broken circuits in landing light wiring circuits.

111

This allows you to check the landing light circuits only. This may seem foolish, particularly in the light of item 17 following, but that is the way it is with the FAA. So check for current at the light and its connector; if you have current here, you may repair any damage between it and the light. If there is no current here, to all intents and purposes you are not authorized to proceed back to the landing light switch if the wire goes through a multiple connector. If the connector at the wing root is single, however, you may trace the wire back to the circuit breaker or switch.

Corrosion and broken bits are the usual culprits in the landing light, after blown bulbs, of course. (See the following item; this one does not authorize you to replace the bulb!) Clean out any visible corrosion, make sure contacts are contacting properly, and so forth. If you can get the thing working intermittently, even most of the time, but can't get it to function *all* the time, scrap it. When you need the lights, you may really need them.

(17) Replacing bulbs, reflectors and lenses of position and landing lights.

This is the item that licenses you to change your landing light bulbs in addition to the position lights, but does not consider you competent enough to troubleshoot the position light circuitry. It probably has to do with the fact that the FAA doesn't mind your fiddling with accessories, as long as you don't want to play with primary or operating systems," which they consider the main airplane wiring harness to be.

(18) Replacing wheels and skis where no weight or balance computation is required.

If you have these, you should know what to do. The manual that comes with the skis has complete installation information. The initial installation should be done by an A&P, but once they are installed, changing them with the seasons is up to you, if you care to accept the assignment.

When the wheels are replaced in the summer, take proper care in bearing adjustment, etc. Keep the wheels in a reasonable storage place during the winter. I have seen them being used for wheel chocks!

(19) Replacing of any cowling not requiring the removal of the propeller or disconnection of flight controls.

Removing the cowling is relatively straightforward, where it can be done. Remember to replace any wiring to lights or control cables to cowl flaps (Fig. 5-12), and to make sure all fasteners that hold it on are properly secured. (Bowden cables to cowl flaps don't seem to be considered "flight controls".) If you haven't done it before (or even if you have), a checklist wouldn't be out of order. You will most likely be removing it for additional work on the engine, and by the time you get to putting it back on, you may forget something. Taking it off again to hook up a forgotten landing light wire can be boring; having it fly off in your face when taking off is all *too* exciting.

ENGINE & OPERATING SYSTEMS

(20) Replacing or cleaning spark plugs and setting spark plug cap clearance.

Tools: A deep-socket spark plug wrench, ⅞-in; necessary extension and a torque wrench.

Spark plugs are one of the major indicators of engine condition. From this point of view it is a good idea to check them yourself, although it is a relatively difficult procedure. It also requires a spark plug cleaning machine. You have to break my cardinal rule at the beginning of this chapter: "Don't borrow tools." There is a good alternative—you remove and replace the spark plugs and have your A&P clean and gap them. This way, you can also get a second

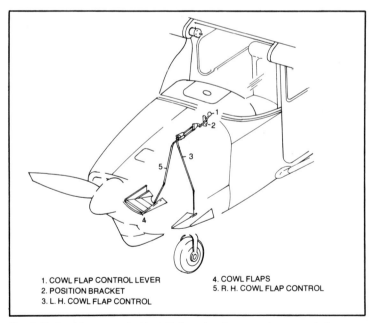

1. COWL FLAP CONTROL LEVER	4. COWL FLAPS
2. POSITION BRACKET	5. R. H. COWL FLAP CONTROL
3. L. H. COWL FLAP CONTROL	

Fig. 5-12. Cowl flaps, a typical installation, showing operating controls that must be removed to remove cowl.

opinion on the engine condition, as well as give him some business in exchange for all the information he has been giving you when you run out of knowledge (Figs 5-13 through 5-18).

Removing spark plugs requires a deep socket of the right size, usually ⅞-inch hefty extensions, and handles that will allow a lot of leverage, as plugs tend to get sticky. Sometimes, universal joints or other accessories will be necessary to get the spark plugs out in a particular installation. If your tool box is not too generously equipped, the first time you remove them, it is best to do it on a day when you can run off and complete your tool kit after you know what you will need.

As the spark plugs are removed, they should be carefully segregated as to cylinder and position, so any problems indicated by the spark plug can be identified as to cylinder. (An egg carton is good for this.) Cleaning the plugs should be done by hand, checking them and finishing up on the machine. If you have a good enough relationship with an A&P to borrow his machine, it is probably good enough to have him show you just how to use it. At the same time, he can advise you about your spark plug and engine condition.

After the spark plug has been cleaned and gapped, it should be carefully cleaned (externally) in solvent (*not* gasoline), replaced in the engine with a new gasket, and torqued to the recommended specification. If you want to put anti-seize compound on the threads, be very careful not to let any get near the electrodes. It will foul the spark plug quicker than you can wink, and require redoing all the work. A very small amount placed on the second thread of the spark plug will lubricate sufficiently, while eliminating the possibility of fouling the electrodes. The leads and "cigarettes" should be carefully cleaned and replaced—when replacing the "cigarettes," make sure they are an easy fit. Don't jam them in.

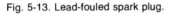
Fig. 5-13. Lead-fouled spark plug.

Fig. 5-14. Carbon-fouled spark plug.

(21) Replacing any hose connection except hydraulic hoses.

This item basically applies to the vacuum and instrument hoses. If you fly instruments, you know how much you rely on these.

Instrument vacuum hoses come in two types, one prefabricated with AN fittings on each end; the other, more modern sort, is plastic held over the fitting with a ratchet-type clamp. The AN, prefabricated type are replaced by part number, the plastic type are cut off stock to the proper length. The ratchet clip on this type hose is removed by slipping a small screwdriver point under the ratchet tongue and lifting it. The clamp should release. Frequently it doesn't, and the clamp must be destroyed to remove it, so don't try this without a supply of these clamps on hand. They are replaced by sliding the ratchet into position under the tongue and compressing the clamp with a pair of pliers. The hose used *must* be the right sort of hose.

Engine hoses for the pressure gauges are low pressure hoses and ordered by part number. Where they pass through the firewall

Fig. 5-15. Spark plug gap erosion.

115

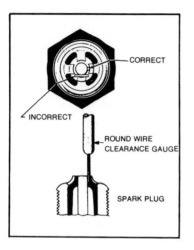

Fig. 5-16. Proper use of measuring gauge with aircraft spark plug.

or are tied down, it is very important to make sure all the grommets, sealing devices, and sealing compounds are replaced properly to avoid dangerous leaks and the possibility of the hose abrading.

These sorts of hoses should, in my opinion, be left to an A&P. The hose work seems simple, but requires the sort of care and expertise that many of us do not possess. Even if the hose work is properly done, there is always the very real possibility that something vital behind the panel will get bashed, bent or broken while replacing the hoses, only to cause trouble later (Figs. 5-19, 5-20).

Fig. 5-17. A damaged "cigarette," from careless removal or installation. Be careful.

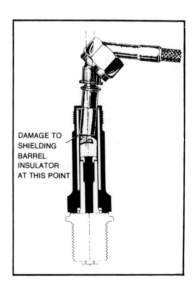

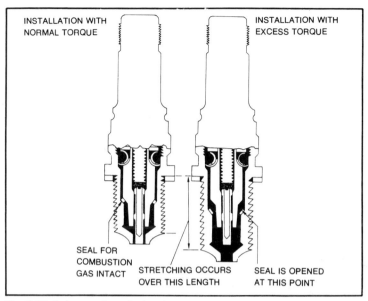

INSTALLATION WITH NORMAL TORQUE

INSTALLATION WITH EXCESS TORQUE

SEAL FOR COMBUSTION GAS INTACT

STRETCHING OCCURS OVER THIS LENGTH

SEAL IS OPENED AT THIS POINT

Fig. 5-18. A damaged spark plug due to over-tightening. Use a torque wrench.

(22) Replacing prefabricated fuel lines.

These are straightforward. If the line is getting soft, or shows any other sign of wear or leakage, it should be replaced. The replacement is ordered by part number. When the work is completed, the line should be checked for leaks. Turn on the auxiliary pump. If the airplane has no electric pump, turn the fuel "on." Leave the pump on, or the fuel selector on long enough to be sure the hose and its fittings are not leaking. *No job of this nature is complete until it has been tested.*

(23) Cleaning oil and fuel strainers.

This is one of the more antique items in Part 43, probably dating back to the days when a pilot had to clean out the fuel and oil system before every flight (and frequently in the middle of one). Some of the work described here has changed dramatically since this paragraph was written, and is beyond the scope of the average pilot, but the fact remains that it is legal for the owner/pilot to do it.

The fuel filter bowl is relatively easy (Fig. 5-21). Turn off the fuel, drain the strainer bowl, clip the safety wire, and remove the bowl and the components therein. Clean all of these, including the gauze screen, in acetone or carburetor cleaner, and reassemble, using a new gasket. This is absolutely necessary: old gaskets almost always leak. The retainer should then be tightened to

117

specification and the fuel turned on to check for leaks. *The job is not complete until it is checked for leaks.* In the absence of leaks, the retainer should be safety wired, paying attention to the instructions under item 5.

The carburetor screen also comes under this paragraph, and it is a different story from the fuel strainer. Incorrect reassembly of this component can junk your carburetor and cause leaks, fires, and other serious problems.

If you must do this yourself, have your A&P show you how it is done on your particular carburetor or fuel-injection system. When reassembling this screen, use new gaskets and be extremely careful in your leak check. If there are leaks, do not try to eliminate them by excess tightening of the nut. This invariably leads to stripping the threads in the carburetor housing, ruining the carburetor. If there is a leak, it is almost always in the gasket or gasket seat on the carburetor housing.

The oil screen is removed (as recommended by the engine maintenance manual) and carefully inspected for foreign material. If

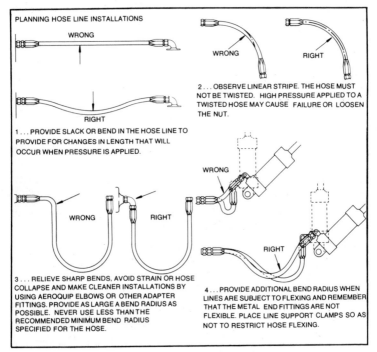

Fig. 5-19. Proper installation of flexible hoses. Take care with these and check them when inspecting the engine, changing oil, etc.

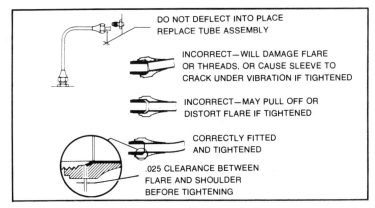

DO NOT DEFLECT INTO PLACE
REPLACE TUBE ASSEMBLY

INCORRECT—WILL DAMAGE FLARE
OR THREADS, OR CAUSE SLEEVE TO
CRACK UNDER VIBRATION IF TIGHTENED

INCORRECT—MAY PULL OFF OR
DISTORT FLARE IF TIGHTENED

CORRECTLY FITTED
AND TIGHTENED

.025 CLEARANCE BETWEEN
FLARE AND SHOULDER
BEFORE TIGHTENING

Fig. 5-20. Don't mangle flared fittings. They only fit one way.

there is any present, show it to your A&P; he will analyze it or send
it away to be analyzed. Foreign material may be normal under
certain conditions, but it should be shown to your A&P for a
determination as metal in the oil can be the first sign of major engine
difficulties. The screen should be cleaned and replaced, with new
gaskets. It should be leak-tested by running the engine for a reasonable period, and carefully safety wired as recommended in item 5
(Fig. 5-22).

The newer, automotive-style filters require a special tool for
their removal and installation. Once the filter is removed, it will
have to be cut open and inspected for foreign material, which is a
messy job at best. If you find any, clear it with your A&P before
operating the engine. When the filter has been replaced, it should be
leak-tested.

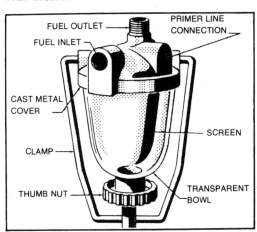

FUEL OUTLET
FUEL INLET
PRIMER LINE
CONNECTION

CAST METAL
COVER

SCREEN

CLAMP

THUMB NUT
TRANSPARENT
BOWL

Fig. 5-21. A typical
fuel strainer.

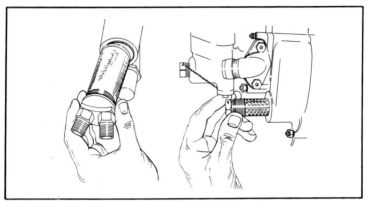

Fig. 5-22. Oil pressure and scavenge oil assembly.

If you plan to do the work on your oil system, remember that *some* leakage from the engine is inevitable, so have a pan handy to catch it.

(24) Replacing batteries and checking fluid levels and specific gravity

Efforts put into good battery care repay themselves many times over, both in length of service and reliability (safety). You should have an intimate acquaintance with your battery, its terminals, and battery box.

The battery itself is much like the one in your automobile, and it suffers from the same complaints. The electrolyte should be maintained at the correct level by topping up with distilled water. Drinking water frequently has a mineral content high enough to cause damage in your battery, so it should be avoided. Don't have too much faith in the little plastic "distilled water" bottle at your FBO. More frequently than not, the water comes from the tap in the men's room.

If the battery suddenly starts using more water, or shows signs of spattering from the filler caps, it is a sign of overcharging. Continuous overcharging will kill your battery quickly, in addition to being a sign that something is wrong with the alternator or generator regulator, so it should be investigated by your A&P.

The most important secret to making your battery last as long as possible is to keep it fully charged. When the battery is not charged, a condition called *sulfation* occurs. Enough sulfation and the plates short, rendering the battery useless. When the battery is fully charged, all the sulfur is in solution, and sulfation cannot occur. If you are starting your engine to check for oil leaks, or to troubleshoot the electronics, run it long enough to charge the battery fully before you shut it down. If you are using the battery to check avionics or other electrical circuits, charge the battery from an external source. Whatever you do, make sure it is fully charged before shutting down and going home.

Winter is the real battery killer. The engine is more difficult to turn over, while at the same time, the output of the battery drops radically as the temperature goes down. Just when it has to work hardest, it has the least to work with. In addition, if the battery is left in a discharged state, it loses its "anti-freeze" qualities, and the battery freezes.

When the weather is *really* cold, preheating the engine and an external battery start are very good ideas. They save both the battery and the engine from the cold weather.

The condition of the charge of the battery can be determined with a hydrometer (Fig. 5-23). If, after charging for a sufficient period, or until after the ammeter comes back to neutral showing a full charge, the battery will not show a full charge on a hydrometer, it is defective; demote it to duty on your lawnmower, and put a fresh one in your airplane.

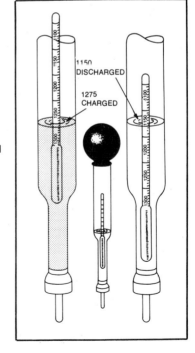

Fig. 5-23. A hydrometer, showing full charge and discharged readings.

Quick charges are to be avoided. They overheat the battery, which is detrimental to its life, and in the worst case, they can warp the plates and destroy a battery. If the battery is low, trickle charge it overnight, or start the engine and charge the battery with the generator or alternator. You will save money in the long run.

The battery should be kept clean, and the terminals covered with Vaseline to help avoid the forming of corrosion at the terminals. The battery box should be flushed with water and baking soda until all acid is gone. This will be indicated when the solution stops foaming. Do not use baking soda on the battery itself, as a very small amount in the battery will ruin it. Once the battery box and hold-downs are clean, they should be painted with an acidproof paint.

When removing the battery from the airplane, be sure to disconnect the ground connection first, then the hot side. This will eliminate sparking and possible explosion from the ignition of the battery gases, in addition to eliminating the possibility of frying the diodes in an alternator-equipped aircraft—an expensive proposition, at best.

Chapter 6

Choosing an Airplane

As, through the years, we have "traded up" in the equipment we fly, I have noticed a singular phenomenon of the aviation world. Nobody seems to care or is in the least impressed by what you fly, other than your banker or yourself. You may be able to cut a corner here and there and scrape up the payments for a Mercedes, but you can't do that for your Learjet.

As we progressed up the aviation ladder of sophistication (and expense), I was surprised when the line boys weren't more impressed with the Cessna 182 than they had been with the 172. (Most of them, I am sure, didn't know the difference.) The 182 RG made no impression, and our current move to a light twin has had exactly the same effect. As a matter of fact, the Learjet that shakes up even the most jaded lineboy in Virginia has no effect whatsoever at National in Washington, or Kennedy in New York. There, the ante is higher—to walk off your airplane directly at the gate seems to require at least a Gulfstream II or better—the Citations and such are relegated to the parking lot, and their passengers rub elbows with the single engine set in the bus on the way to the terminal. If you absolutely *must* be the center of attention wherever you go, start saving up for a Beech Staggerwing or a P-51.

It is very simply futile to trade beyond your *real* needs and financial ability in the aviation world. The ultimate is beyond the reach. If your usual trip is 300 miles, with a once yearly 600-mile jaunt with two people, a Bonanza is not the answer to that problem. More likely a Cessna 172 is, although the Bonanza may fit your flight qualifications and self-image better.

This chapter may help give you a real, *practical* idea of what you need in an airplane. If you are one of those people who think that practicality has no place in the aviation world, you are probably right with regard to a certain type of owner. I can think of many happy pilots who have Dukes and upwards sitting in hangars, used on 200-mile jaunts twice a month—airplanes that wear out sitting rather than flying. On the other hand, most of those owners are not reading this book—they have the money to play the game, and they don't have to count beans.

The trade-up game, financial version, doesn't affect the aviation world the way it does the automotive. A more insidious version of the game does, however. The name of our game is Qualifications and Ratings. To the extent that an instrument or other advanced rating makes us a better pilot, more power to it, but there seems to be a prevalent opinion that an instrument, multi-engine rated pilot shouldn't be seen in anything less than an airplane that reflects his qualifications. If you are flying a light twin, obviously a lot of the refresher work with your instrument rating, landings, engine-out drills and the like must be done with the airplane you normally fly. If, on the other hand, you find you are having a bit of difficulty with the charts, or keeping the needles centered, you would be financially a lot better off in a trainer, either airplane or simulator.

FIGURING COSTS

There are many ways to figure the cost of operating your airplane, or another model. The lists in this chapter will help you do that down to the penny, tight as Scrooge, and you will probably have to hire Bob Cratchit to do the figuring if you want to figure the cost of more than one model for purposes of comparison. If you are as adept as figures as I am, you are guaranteed to like my revolutionary, throw-away-your-pencil-and-computer method of figuring approximate relative operating costs for almost all current models of airplane.

Depending upon whether you want to have fun or save money, this approach requires the use of an airplane, or, for the dull and admirably frugal, the telephone. Get in your airplane (or on the telephone) and get the hourly rental rates of the airplane you are interested in. Try to get figures from as many dealers as possible for each type. You will find some that are obviously not competitive; these should be discounted. Make sure prices quoted are for recent models. You want the price to reflect *current* costs of operation.

The point is that current rental fees reflect accurately the true total cost of operating the airplane. They include the obvious as well as the less obvious (and often forgotten) costs. On the whole, the figures obtained will be more useful as comparative figures rather than absolute ones, but you can generally figure a one-third markup and be pretty close to the actual cost.

The prices will reflect true costs, because the FBO whose charges are too high or too low will not stay in business long. This assumes reasonable competition in the aircraft rental business in your area. Do not assume that high rental fees in your area, as opposed to other areas, are always a sign of lack of competition. The high rental fees charged in your area most probably reflect a hidden cost of doing business there, one which you might do well to figure into your cost formula.

Remember that these cost figures require reasonable use of the aircraft. If you are like most general aviation pilots, you probably shouldn't really figure your costs by the hour, but rather by the year. The cost by the hour is of more interest to the fleet operator, but few general aviation owner/pilots fly frequently enough to make these figures worth anything. You figure your automotive costs by the year; why not your airplane?

FIGURING NEEDS

The next place to go is your logbook. When deciding what kind of airplane you need, you need facts and figures. Otherwise it is too easy to confuse *wants* with *needs*. The logbook figures should be entered in the chart (Table 6-1), breaking down all trips into mileage groups, and training into aircraft familiarization, flying around the pattern for the fun (and experience of it), and training that could be done in a training aircraft or simulator.

Follow the instructions on the chart, and the result will be the number of hours you actually flew last year. This figure must then be adjusted to take into consideration the additional flying that you plan to do in the new airplane.

The trip mileage groups will tell you the most about your needs. If your flying is mostly in this category, it should be easy to choose the practical airplane for this category. If your trips are mostly under 200 miles, and your flying is predominantly in column

A, the practical choice should be equally obvious. Don't cheat; if your psyche needs a Bonanza, there will be room for that later. The purpose of this exercise is to take the calculated, corporate approach to choose, on an entirely practical basis, the airplane you need.

Once the chart is filled out, look at your trips (Table 6-2). Under each of the three categories, put an airplane that, in your opinion, best suits the needs of these trips; i.e., the one that you would rent for each category of trip, if you had unlimited availability of rented aircraft. Now you have a good idea of what sort of airplane you would buy if you were a cost-conscious corporation looking for the maximum return on your investment. If the bulk of your flying is in the 200-mile category with a yearly 1200-mile trip to visit your in-laws, it would probably be more cost-effective to *rent* a faster, more sophisticated airplane for that trip and settle for a less-impressive airplane for your regular use.

Fortunately, we do not all have to think like a cost-conscious corporation and you will have to factor your desire quotient into the equation. You will notice there is no room for it in the chart, but that's what charts are for. If you want a Cessna 195 or a Beech Baron, a Yankee will not fill the need or the hole in your heart. If you had to fly it, you might as well give up flying. You should also bear in

Table 6-1. Aircraft Selection Chart.

| | LOG BOOK HOURS | | | TRIPS COL "C" | | |
	A	B	C	200 mi.	400 mi.	600 mi.+
PAST YEAR						
ANTICIPATED WITH NEW AIRPLANE						
TYPE AIRPLANE CONSIDERED BEST FOR EACH USE						

A. FAMILIARIZATION WITH AND CURRENCY WITH AIRPLANE, MAINTENANCE

B. AROUND THE PATTERN, SHORT HOPS, ETC.

C. TRIPS, BREAKDOWN BY MILEAGE

Table 6-2. Chart, Filled Out Example.

	LOG BOOK HOURS			TRIPS COL "C"		
	A	B	C	200 mi.	400 mi.	600 mi.+
PAST YEAR	50	60	130	10	20	100
ANTICIPATED WITH NEW AIRPLANE	50	60	300	10	20	400
TYPE AIRPLANE CONSIDERED BEST FOR EACH USE	N/A	PIPER "CUB"	BONANZA	PIPER CHEROKEE 140	CESSNA 182 RG	BONANZA

mind the expense of staying current in the airplane you need to rent for your occasional trips!

LEMONS

When choosing an airplane, new or used, it is important to know which ones are, or are thought to be, lemons. Airplanes become lemons for different reasons. It is important to understand the process, for almost all airplanes are affected to at least some degree by the process.

Bummers

Bummers have engine or airframe problems that have never been successfully or permanently fixed. This sort of airplane should be avoided *at all costs*. They are generally bought by people who are not aware of the problem the particular model has, and when they become aware of it, have to take the loss inherent in the sale of this sort of airplane. Bummers should not be bought under any circumstances! Replacing an unreliable engine with a more reliable one may make the airplane satisfactory for your needs, but the airplane will still be a lemon when you try to sell it, and generally the cost of the engine modification will not be recoverable.

Orphans

Some of the most interesting airplanes are orphans. In fact some orphans are not orphans at all. Many early, out-of-production makes have better parts availability than their cousins manufactured by companies still in business. The orphan classification has to apply to airplanes manufactured by weak manufacturers on the brink of extinction as well as limited production and foreign aircraft, or anything not made by the big three, or three and a-half.

Orphans should be financially calculated as shrewdly as possible under the circumstances. They generally have something other airplanes don't, traded off against something the others have but they don't. They are often underpriced and have to be sold that way. The airplane that has lots of room inside but is slow and thirsty will always be the same, and valued by the market at the same relative value. Like an undervalued stock that nobody appreciates the true value of, the undervalued airplane is unlikely to be suddenly re-evaluated to your profit in the market place.

There are, of course, exceptions to this rule. Fast, fuel-efficient aircraft are, as of this writing, being re-evaluated in the light of rapidly rising fuel costs. On the whole, however, don't count on it. Airplanes in this category are most likely to depreciate at a slightly higher rate than their more successful brothers. If you want an airplane that is roomier and slower, or one that is thinner and faster, by all means consider this category. The financial considerations are not too difficult, and there is no question that there can be pride and pleasure in owning something out of the ordinary. At the same time it should be remembered that the market for this sort of airplane is limited. The ordinary, garden variety medium-priced aircraft of the big three will sell quickly, while it may take some time to get a reasonable price for your interesting orphan when you want to sell it.

Unsuccessful Models

The major manufacturers all have models that are not selling. Some of these were once big sellers, but are not any more; others may *never* have been very successful or desirable. You would do well before considering one of these to find out *why* the model has not been terribly successful. Sometimes, on intitial introduction the airplane got a bad reputation. The early problems were corrected, but the airplane never did catch on. Others just missed finding a niche. Some of these are good buys, and others are bummers. Avoid the bummers and carefully weigh the others, treating them like orphans, and you won't be hurt. When buying an airplane in this category new, be particularly careful of the price you pay. If you do not strike a good deal, the initial depreciation can be very high. It is very hard to sell something nobody else wants.

Poor Performers

Most training aircraft fit into this category. They simply do not go far or fast enough, nor do they carry enough to have a good market. Add to this the fact that there is a constant availability of these airplanes as the flight schools turn them over, and it explains why their prices are generally weak. If you want an around-the-pattern airplane to play with, an older classic or semi-classic trainer is probably a better buy.

Some airplanes were never meant to be slow, but somehow turned out to be that way. The same holds true for carrying capacity, landing characteristics, stall speeds, fuel consumption, and so forth. These are also performance-blighted, and should be carefully weighed before buying. Their reputations are unlikely to get better.

GETTING THE NEWS

The general aviation press seems to have a certain reticence to discuss problems with avionics and airplanes. The mass-circulation, four-color magazines are a good place to work up an interest in aviation during icy winters and thundery summers, but seemingly (in most) nothing manufactured by an advertiser may be criticized. In short, it is the upbeat, positive attitude that sells aircraft, avionics and magazines. Pollyanna, unfortunately, does not give the potential purchaser of aircraft or avionics the information necessary to invest his money.

If you restrict your aviation reading to these magazines, how do you find out about the airframe that requires constant inspection and repair, or the engine that has a major problem? Who will tell you about the avionics that simply don't work? An airplane that suffers from such defects will be difficult to operate and harder to sell. The potential financial loss is great, and the aggravation of owning such a thing has turned more than one aviation enthusiast to less expensive and burdensome hobbies, like collecting Ferraris.

The negative information you need to intelligently judge your investment in an airplane is not as readily available as the positive, upbeat sort. Newstands neither carry nor display it, nor will your local aircraft dealer volunteer just where the problem areas are in the airplanes he sells, although he may be glad to fill you in on the weaknesses of the competition. In fact, this is one of the places where you *can* do some detective work, taking this sort of information with a grain of salt, of course.

Once you have decided that you like a certain airplane, talk to people about it. If you have access to people in the aviation business, talk to them. Mechanics, owners, airport bums, and even dealers can fill you in. For some models (particularly the older ones), there are clubs that can be very helpful. Try to find a mechanic who specializes in your airplane who doesn't work for a dealer. He may be somewhat less reticent than a captive mechanic, but I have found all mechanics are pretty straightforward about problems that an airplane may or may not have.

While you are talking to anybody you can about your potential purchase, send away to your old and constant friend the FAA for their information. Never buy an airplane until you have read all that the FAA has to say about it.

The FAA speaks through Service Difficulty Reports (SDRs) and Airworthiness Directives (ADs). Neither one of these is to be missed before deciding whether to buy a particular model of airplane, new or used. Service Difficulty Reports are compiled from the Malfunction or Defect Reports (MDRs) which are received by the FAA from pilots (voluntarily), repair stations, and air taxi operators (required). These reports are fed into a computer which analyzes them and groups them together in categories of repairs and types of airplanes. If there is an excess of reports in any area, the problem is discussed with the manufacturer, and may result in a Service Bulletin from the manufacturer or an AD from the FAA.

Although these reports can be heavy reading, they are a *must* for those contemplating buying either a new or used airplane, as they point out troubles in a particular airplane before they become ADs, thus helping you avoid unpleasant and expensive surprises. They will also indicate, for instance, sudden difficulties experienced with an engine change or other design modification in a previously trouble-free older model, allowing you to avoid the problem before buying the airplane. SDRs are available from the FAA for the particular model you are interested in; the records are for five years only.

FAA
Safety Data Branch
Oklahoma City, Oklahoma 73125
(405) 686-4171

The price for SDRs varies according to the quantities of same. The more difficulty reports, the more they cost—perhaps an omen of things to come!

ADs of course must be complied with, as they are FARs, ammendments to Part 39, and an airplane cannot legally be operated when it is not in compliance. A record of compliance should be kept in the airplane logbook, but it doesn't hurt to doublecheck by getting the ADs on the model you intend to buy and checking them for compliance, as well as for any current ADs that will need expensive compliance in the near future. Lists of ADs are not available from the FAA, as they are published in the Federal Register, but they are available from:

Aircraft Technical Publishers Inc.
655 Fourth Street
San Francisco, Ca 94107

who keep them in a computer. They are accessed by N number, so you get only the ADs that apply to your particular airplane. A word to the wise: you do not actually get the whole AD, with the instructions for compliance, etc., but then only a mechanic really needs to see the whole thing. Frequently, the best person with whom to discuss ADs on a particular model is a mechanic who specializes in that airplane.

Both of the above publications are available from the Aircraft Owners and Pilots Association's Aircraft and Airmen's Records Department, where they cost more than from the sources directly, but are available quickly and simply, to both members and non-members of the organization. This information represents only a small part of the services that the AOPA renders its members in the area of airplane buying. This is one area in which they really shine. Their address is:

AOPA
7315 Wisconsin Avenue
Bethesda, Md. 20814
(301) 654-0500

The *Aviation Consumer* is a more aggressive version of Consumer Reports. Aimed at the general aviation market, they look for warts in products and take an ill-concealed delight in exposing them to public view. They cover particularly aircraft and avionics, spending most of their time where you put most of your money. In my view, you cannot afford to buy an airplane or avionics without consulting them. Their address:

Aviation Consumer
Subscription Department
Box 972
Farmingdale, N.Y. 11737

Trade-a-Plane should be the bible of the airplane buyer. It is the bible of everybody in the business—the New York Stock Exchange for the parts, equipment, and used airplane business. Here you see *real* prices (both new and used), price trends and availability, and can spot troubles with particular models quickly. (Even the dullest among us realize that there must be something wrong with an airplane if the later model sells for less than the older ones.) If you are buying a new airplane, *Trade-a-Plane* can be great help. If you are buying a used one, you cannot do without it.

Trade-a-Plane
Crossville, Tenn. 38555

Chapter 7

Keeping It Clean

The engine is being properly flown and maintained—it will last. The avionics downtime and expense has been severely reduced by pilot troubleshooting and operating techniques. Good preventive maintenance practice, and do-it-yourselfing has kept general maintenance at a very high level, while keeping costs down, but the danged thing is lasting *so* long that it is an embarrassment at every airport.

The paint is dull and chipped. The plexiglas is crazed and discolored. The interior has hardly benefitted by all those years in the hot sun. They never considered the real world of soda pop, melted chocolate bars, and hydraulic fluid—not to mention your little nephew—when they designed that beautiful velvet interior.

Maybe it would have been better to let the maintenance slip a little bit. Taking all that care of the engine doesn't pay off enough at trade-in time if the machine looks doggy. Well, it's too late now. The airplane's like new where it counts, and you'll just have to do something about the rest of it.

The "trade-when-the-ashtrays-are-full" school of financial planning was until recently in vogue in the automobile business, before they got so high, but it never did get very far in the aviation field, in spite of different paint jobs every year in a half-hearted attempt to make you feel out of date. The reason, of course, is the expense of the airplane. It just doesn't, by any amount of figuring, pay on the bottom line. In addition, airplanes don't change very much over the years, nor are they as complicated as automobiles; there is not as much to wear out. Airplanes tend to make up for this by the high maintenance required, but the maintenance tends to be almost the same for a new airplane as an old one. More than one operator can show you how much less it costs to operate his older, de-bugged airplane than it does to operate a current one.

The answer is, of course, preventive maintenance. This includes maintenance to the cosmetic aspects of the airplane—the exterior finish and the interior. This area is almost totally up to the owner/pilot. Few people will do the sort of work outlined in this chapter for you; if you could find someone, it would make the work uneconomical, the labor cost would be too high. There is no reason for a modern airplane to look anything other than brand new after five years of use. Modern finishes, plastics, and upholstery materials can hold up very well indefinitely, given half a chance and minimum care.

CHOOSING A LONG-LIFE AIRPLANE

Like choosing a proper airplane for its anticipated use, efficiency, or need, there is no use in getting too practical in this area. If you want a low-wing, velvet-interior, polished-aluminum-exterior airplane—go ahead. A practical airplane will not fill your needs in ways that probably count more than being practical. More than one corporation has a full-time man on the payroll to do nothing but polish the aluminum exterior of their otherwise rationally-chosen corporate airplane.

Interior

At one end of the scale would be a light-colored velvet, carpets and all, and at the other end a brown plastic. There must be a happy medium somewhere in the practicality scale.

If you are buying a new airplane, you will be surprised at the broad choices that are available. If you *must* have the light grey velvet, how about a more practical, darker color for the carpets? If you want cloth, there are options to velvet that wear better and hide unavoidable accidents better, while still being cool and elegant.

Plastic can be hot, but twenty years from now it will still look much as it did when it left the factory. Not so many years ago, plastic was an extra-cost option; people were tired of the fast wear and grungy look of fabric after a few years. How quickly we forget; now fabric is all the rage. Of course, fabrics are longer-wearing now, stay cleaner, and look fresher than they used to, but the plastics have been improved too. It isn't quite as sticky as it used to be, and if it is pierced, it is reasonably comfortable in all but the hottest climates.

An alternative that is not too often considered these days is the real thing, leather. With proper care, it will last longer than plastic, and is cooler. In fact, it is cooler than many new fabrics, and when the price is considered as part of the whole cost of the airplane, it doesn't seem so unreasonable.

But more than anything else, it is *real*. As it grows older, it gets wrinkly (just as you and I) and as friendly and comfortable as a pair of old shoes. It has character and smells good. Nothing smells as good in the whole world. If you can't tell the difference between leather and plastic, and don't care, don't bother. If you are tempted, read about leather care in the upholstery care section of this chapter and consider the option. It won't change your life, but it may add to it.

There are still some airplanes being built with a minimum of plastic bits and pieces. This makes sense from a practical point of view, as the little plastic pieces are the ones that crack, warp and fade the fastest. Unfortunately, the most expensive airplanes are the ones that have the simplest interiors, with the least amount of plastic bits, so from a cost-benefit point of view you might do just as well getting used to the idea of replacing these as they deteriorate. The prices of the little bits of decorative plastic are astonishing. A little piece of imitation walnut trim can cost enough to buy a whole tree of the real thing.

Some of the things that go into making a decision as to the practicality of a particular airplane are not so obvious as the selection of interior upholstery material, and one of these is the position of the wing. There is no question about it, the high-wing airplane does better here. The high wing shades the cabin and keeps the weather out; in addition, high-wing airplanes have less glass area to let the sun and heat in. Over the years this can make a great deal of difference, but the edge that the high-wing has here may be mitigated with extra care of a low-wing.

As a matter of fact, if you are careful, don't carry dogs or children (uncontrollable variety), and are willing to put some time and effort into care and maintenance, even the most impractical interior can be kept fresh for a long period of time. On the other hand, don't have this sort of interior in an airplane that is on a leaseback—I have seen these look like antiques after a few months!

Exteriors

Years ago, certain colors faded quickly. These were the reds and blues. Nowadays, finishes are much improved and last longer, and manufacturers claim that reds and blues don't fade. This is not strictly true. They still fade somewhat, but they last so long that fading is no longer really a problem. If you keep your airplane seven years, and maintain the exterior finish reasonably well, fading shouldn't be a real problem, but if you want to be really practical stay away from these colors, especially in the darker shades.

If you are buying a new airplane, pay for the urethane paint option, if it is available. Polyurethanes are a giant step forward in the paint business. It will more than pay for itself in reduced maintenance and longer life.

If you come from the era when airplanes had big, round engines that made the right noises, and fuselages that were silver, you (like me) are probably not too old to dream, but you are definitely too old to polish the thing as much as is necessary to keep it looking the way it should.

If you are buying an airplane that is covered in fabric, re-member that light colors not only hold up better, but they also protect the fabric underneath. This is probably academic, because as of this writing there are only two airplanes being built with fabric covers—but if you have a choice, keep it in mind. (Cubs always look better in yellow anyway.)

EXTERIOR FINISH

Years ago, people used to wash, wax, and polish all their posessions. No matter what it was, it was washed, waxed and polished. They still sell waxes and polishes in automobile supply stores, but our hearts are not in it. Waxes and polishes aren't the big sellers they once were—it seems to me that most of the cans that are sold live on shelves of garages, hangars, and barns, helping to weigh the structure down in case of hurricane or tornado.

Part of the reason for this is the improved finishes that are used today. A urethane-finish airplane seems to go on forever, looking reasonably good. Even lacquers and the humble enamels have been so improved that they do reasonably well with minimum care.

Washing

In spite of this, airplanes should be washed; particularly if the airplane is kept outdoors in a coastal area, or one that has industrial smog, soot, or other detrius detrimental to the finish. A hangared airplane needs washing only when dirty, or before waxing, but an airplane parked outside in a hostile environment should be washed off as frequently as possible.

Usually the upper surfaces don't need a very strong detergent. I mix up a very weak solution of dishwashing (liquid) detergent in water, and sponge all the upper surfaces. Do not do this in the hot sun; the finish may streak and spot. Rinse this off well, and wipe it off with a chamois. Mix up a stronger detergent solution and go after the generally dirtier underside.

If necessary, a spray-type detergent such as Fantastik or a similar product can be used to remove oil buildups on the belly of the airplane. This sort of detergent is particularly good for removing exhaust stack deposits. When using this kind of cleaner on the finish, use it as briefly as possible, and follow immediately with a mild detergent wash and rinse, as this sort of detergent is very strong and will harm the finish if left on. I limit its use to the places where nothing else will work. Solvent isn't nearly as hard on the finish, and is better than oil, but should also be washed off as soon as practical after it has done its job. When choosing a solvent, make sure it will not remove the finish. *Do not use any strong detergent or solvent on the windows!*

The strong cleaning should be followed by a regular washing, rinsing and drying. If the airplane isn't absolutely clean on the underside, don't ruin your chamois in the oil. Nobody sees the underside anyway; a fifty-fifty wash job is good enough here. (At fifty feet, and fifty miles an hour, it looks good.) When washing, particularly with a high-pressure hose, avoid any areas or orifices that may be damaged by excess water, particularly static system ports.

Touching It Up

This isn't what you do to airplanes at airshows, but what you do to your airplane before waxing it. (Modern waxes are hardly ever wax any more; they are mostly silicones and other miracle things, but I will continue to call them wax to keep things simple.) The time for touching it up is before waxing, as the silicones in modern waxes interfere with paint bonding. Small scratches, chips, and the like should be touched up with a matchstick (not a brush) dipped into the paint. Catch all the spots around the doors, and you will be surprised what a difference this makes in the looks of the airplane. Larger areas where the paint has chipped in larger pieces should be touched up with a spray can of matching paint. If the area is large enough, make up a mask of cardboard about the shape of the hole, and about half the size. Hold this between the nozzle and the surface to be painted; it will act as a stencil and give you a reasonably good feathering into the existing finish. You can rub the little overspray off with rubbing compound after giving the paint a reasonable time to dry, but don't overdo it; you are not painting an antique Rolls Royce. You will be surprised how good even a relatively rough job will look.

If you do this regularly as you wash the airplane, it will never really get chipped looking, and the amount of work touching up the paint will never get out of hand. While doing this, you can also pay attention to fuel-leak areas and other discolorations. Fuel leaks are getting to be a real problem these days as the amount of ketones in the fuel is increased in an attempt to raise octane ratings cheaply. Some fuels are almost 30 percent ketone. If the name ketone seems vaguely familiar, it should be; most paint thinners and removers are made of ketones. Fuel leak areas may be washed with light detergent, sprayed with spray detergent, or rubbed with rubbing compound depending on how bad the problem is. If it is terminal, give it a touch-up spray; it won't look any worse than it does already.

Once the washing and touch-up is done, it is time to wax. Most of the new waxes are easier to apply than the ones you knew in your youth (even if that was only five years ago), and if you haven't done this sort of thing in some time, you will be pleasantly surprised with the comparative ease. If you choose one with a cleaner in it, the cleaner is usually a rubbing compound which will, if you wax frequently enough, rub through the surface of the paint, causing it to dull. (In extreme cases, you can rub clear through the paint.) If you are an infrequent polisher, this sort is satisfactory, but if you get the bug and become a zealot, choose a wax without compound in it. If you come to an area that needs some compounding, you can use it selectively, saving the rest of the finish. Rubbing compound is available in white (as well as the more usual pink) which is very effective in getting stains out of white paint. Do not rub too hard or long with any rubbing compound, or the paint will come off with the stain.

If you don't fly too often, a silicone dressing will protect the tires and other rubber parts from checking and other atmospheric damage. It makes them look good too. Long before these dressings were available, I got in the habit of using a spray furniture-dusting wax (such as Pledge) on these areas, which certainly made them look better. I have used it on car tires that are now ten years old, and they still look like new. I like Pledge for all rubber, leather and plastic, antennas, fairing strips, etc. It renews the new look and seems to protect them from the atmosphere as well as discoloration. I am very fond of the stuff, and recommend it for any area where you don't want to put wax.

Acrylic plastics as used in windows and windshields should be carefully washed, with a lot of water, to avoid scratching. They should never be wiped with paper towels to clean them. Keep line boys away from your windshield if at all possible, even if they have the very latest super acrylic cleaning solution in their greasy paws. Unless it is the utmost emergency, you should limit yourself to the recommended cleaning procedure. The surface should be washed with a mild soap and lots of water. The water is the secret; it will remove abrasive particles before the chamois gets there. Bugs are best soaked off. Even the toughest bug will disappear with a long-enough detergent soaking. Occasionally, though, you will run across a bug who, intent on committing suicide, also puts a dent in the windshield.

I feel so strongly about using a chamois for this purpose that I carry one when traveling in a small plastic bucket. This gives the pilot a complete window cleaning kit.

If you have light scratches in acrylic, they may be removed by judicious polishing with one of the rubbing compounds sold for this purpose. Some manufacturers recommend regular rubbing compounds as an alternative, but these are too rough to do a really fine job. If you have a slightly clouded spot left from this sort of compound, it can be easily removed with a finer compound, of the sort recommended.

Heavier scratches may be removed with a 350 wet-or-dry sand paper, used wet on a block. Once the scratch is removed, change the paper to 450 wet-or-dry to clean up the scratches left by the 350. The scratches left by the 450 are removed with 600. All of these should be liberally wetted while being used. The scratches left by the 600, are, of course, removed by rubbing compound.

INTERIOR CARE

Interiors of new airplanes seem destined to impress the new (or prospective) airplane owner and few others. Shortly after the airplane has been weaned from straight mineral oil, the interior is on its way to the last roundup, for life at 7000 feet can be dangerous. Spilled drinks, chewing gum, and oily fingerprints are hard on even the most state-of-the-art hydrocarbon interior. (That melted chocolate bar in the map pocket doesn't help things much, either.) The elusive Nauga, who gave its life so that his hyde could grace the interior of your cabin, deserves a better end.

Messiness in the cabin, to a point, is up to you, although it can be dangerous—organization pays off when things start going wrong. Charts and old flight plans all over the cabin tend to get in the way, and a chart over your avionics cooling vent can cause the most expensive sort of grief. But none of these things has much to do with the long life and subsequent value of the interior.

A chart on the floor won't permanently damage anything, but a melting candy bar will, as will soft drink stains, and greasy smudges from food or the airplane itself. Chewing gum, coffee, and airsickness mess don't help either. The real secret to managing these stains is to remove them as soon as possible. Materials for a quick cleanup should be kept in the airplane, where they are available in flight (if possible) at all times. The longer the mess has to sink into that expensive upholstery, the harder it will be to remove.

To this end, a cleanup kit is a good idea. Keep some alcohol-type window spray (*not* for the windows) in your plastic bucket, along with a wet sponge and the chamois for the windows. If the accident has been large—a spilled drink bottle or an airsickness problem that got out of hand—clean it up as well as you can when you stop for refueling.

Plastic

Plastics may have their disadvantages, but difficult maintenance is not one of them. All the plastic surfaces in the interior respond well to a wrung-out chamois or damp towel. Dishwashing detergent, as you would use for the exterior, is adequate for removing most dirt and smudges. If you have a stubborn smudge or stain, a spray detergent such as Fantastik, used carefully and sparingly, will frequently remove it. This sort of detergent is very strong, and should not be allowed to remain on the surface for long nor be allowed to drip down, as it will streak the material. (It must *not* be used on windows.)

For small spots, it is preferable to spray some on a towel and rub this on the spot to avoid possible problems. After using this sort of detergent, wipe the material thoroughly with a wet towel to remove all the detergent. It may then be sprayed with Pledge. This seems to renew the plastic and keep the dirt off. An occasional cleaning of the plastic can be accomplished by spraying it on and wiping it off, as you would on furniture.

Leather

If you are lucky enough to have a leather interior, its care is also relatively simple. You will find that the side panels and a large part of the trim is, in fact, plastic. It should be treated as recommended in the preceding section.

The leather itself should be treated more gently. Leather is absorbent, so speed in reaching it after an accident is important if you are to avoid stained areas. In fact, leather is extremely difficult to remove stains from, so the best defense is to avoid them. When all else fails try a spray detergent, but *very* carefully. Remove all traces of the detergent. If the spot is still there, it can be dyed over. If the piece of seat that is stained can be easily removed, take it to a shoemaker, or try to find some dye to match from your shoemaker; they generally have a good stock of dyes in various colors. These dyes are more like a paint than a stain, and are quite effective in covering stains and spots.

Leather responds very well to my old favorite, Pledge; it cleans it, protects it from stains, and helps keep it from drying out between treatments.

Treatments? Treatments are the difference between leather and plastic. Leather dries out, cracks, and tears with age. This can be avoided by treating it occasionally with a leather dressing. The period will vary, depending on whether your airplane is stored in an hangar or not, and how much heat the interior is subjected to. Yearly, for hangared airplanes, and three times a year for airplanes under the worst conditions is reasonable.

Recommending a product for this function is very dangerous. Each type of product has its supporters and detractors. It isn't that one product is good and the other better, but that one product is good and the other rots the leather, stitching or backing of the upholstery. I have had "experts" so describe all the products and categories of products listed below.

The product that can be unhesitatingly recommended, one that I have used for a number of years and has always been satisfactory is manufactured by Connaly, in England. Connaly manufactures most of the leather used in automobiles these days. It is called Hide-Food and is available from Rolls-Royce dealers. (It has an R.R. part number). The price is not as unreasonable as the source implies. What makes it so desirable to me, other than the fact that it works, is that it is manufactured by people who make leather, so when somebody tells me that it rots leather, stitching, or backing, I feel I can afford not to pay attention.

In fact, saddle soap, neatsfoot oil (used sparingly), clear wax (like shoe polish), and a dozen proprietary products are useful for periodic leather treatment. The important thing is to do it even when the leather looks brand new. Once the leather starts cracking, it is too late—there is no cure for the cracks, particularly if they have gone through the leather. Immediate care will halt the progressive deterioration, but nothing will really repair them.

Cloth

Cloth interiors require constant care to keep looking even reasonably good. With proper application and care it can be done, but after five years, there is no way that cloth will look as good as even moderately well cared for plastic or leather.

Most modern upholstery materials are made of synthetic fabrics, and therefore are very resistant to dirt and staining, but even if spots are avoided, cloth upholstery gets dirty and grimy with age. A general shampooing with a foam-type upholstery cleaner will help the long-term grime problem, but not eliminate it. Spots are best cleaned off immediately. Greasy spots respond to cleaning fluid; others, water and detergent. If possible, flooding with water can help. I have found the methods in all the home cleaning books to be of doubtful value. Once a spot is set, forget it.

The best thing for cloth material is preventive action in the form of Scotchguard or a similar product put on *before* the problem occurs. The usual end to cloth upholstery is a reupholstery job before the airplane is sold.

Carpets

Carpets are treated much like cloth upholstery. If you are buying a new airplane, I will repeat my advice to get dark carpets, even if the interior if light; they will look much better much longer.

Scotchguard or a similar product helps as much as it does upholstery. A good vacuuming occasionally is another boon to the life of your carpets. Getting a vacuum (and electricity to operate it out to the airplane) is always a troublesome chore, but it is well worth it.

Miscellaneous Parts

Interior chrome, metal and plastic used as trim and for the instrument panel and controls is easily cleaned with an alcohol-based window cleaner such as Windex or a similar product. Avoid getting the delicate area of the instrument panel wet with excess spray. The instrument faces and other very delicate areas should be cleaned by spraying the cleaner on a towel, and then cleaning the glass, rather than wetting the instrument face. Window cleaner is particularly useful for removing nicotine from the aircraft's interior. Never use it on the airplane's acrylic "glass."

Chapter 8

Odds and Ends

In all discussions in this book, I have tried to avoid the obvious. If you fly an airplane into the ground, or blow it up by smoking around the open gasoline tanks, it will cease to have any significant value to either you or a subsequent purchaser, who most probably will buy it from your estate. Certainly, none of us would do this sort of thing intentionally—it costs too much, and the risks are rather higher than most of us would care to assume. On the other hand, many pilots *do* destructive and expensive things to their airplanes without thinking about it. Most of these things are fairly obvious, but deserve a mention anyway.

This chapter contains such items, along with a description of accessories, minor systems, and a discussion on hangaring your airplane—in short, odds and ends.

LANDING GEAR

The primary purpose of the airplane is to fly, but like the water that you never miss until the well runs dry, the landing gear has an important function that you don't think about until there is a mal-function.

Tires

Tires should be frequently inspected for tread wear, ul-traviolet and ozone damage, flat spots, cuts, and bruises. Most of this damage can be repaired if it is done before it gets out of hand.

The kind of operations an airplane is subject to will have a significant effect on the life of your tires, but even an airplane used under the worst circumstances can have a reasonable tire life it the tires are properly maintained.

Look at your airplane's tires. Compared to the ones on your automobile, it is a compromised design. It is more the size of those on your wheelbarrow, designed for lightness rather than strength, yet it is expected to accelerate to 80 miles an hour instantaneously, and then haul a 3000 pound airplane to a stop in 1000 feet. At the same time it is doing all this, it has had 3000 pounds dropped on it, frequently more than once per landing!

Tire pressures are, invariably, the most neglected aspect of aircraft maintenance, as a check at any airport will confirm. The tires work best and have the longest life if tire pressures are maintained to the recommended pressures. Underinflation causes accelerated tread wear which is expensive, and overheating of the sidewalls, which can be dangerous, because it cannot be seen except in the most extreme cases. Underinflated tires also do not allow the brakes to function properly, and can cause pulling and grabbing which may cause a ground loop in a severe case. A blown tire from overheating caused by underinflation can also cause a ground loop, which (in addition to any other damage it may cause to the airframe) almost invariably results in a damaged wheel.

Treat your brakes as if you didn't have them for normal landings. In this sort of landing there is seldom any reason for applying them hard, except to turn off the runway more quickly. If this isn't necessary, don't do it; your tires and brakes will repay you with vastly extended life.

A word here about recaps, which seem to be more than adequate for most uses where tread wear is a problem, if they are of reasonably good quality. They are good, if tread wear is your major problem, but if sidewall deterioration is your problem, brand-new tires are probably a better bet, as the sidewalls of the recaps seem to deteriorate very quickly.

Some brakes are, like engines, more equal than others, but unlike engines, replacement kits are generally available for them at reasonable cost. Usually, the price of the switch will not be fully recovered at trade-in time, but the savings in linings and other parts, as well as their hassle-free operation and superior performance, can frequently make them a worthwhile investment.

In any case, even your less-than-desirable brakes can be made to last longer and perform better by using them carefully. Like the tires, they will not take the strain and abuse the ones on your automobile will. Anything you do to save the tires from wear in the operation of your airplane will also save the brakes, a double return on your investment in careful operation.

Struts

The oleo struts should be maintained as recommended in the airplane's manual, and kept properly filled and inflated. Inspect the dust covers as you inspect the tires. If these get damaged, they can let dirt into the hydraulic strut, which will sharply reduce their service life. After a hard landing, make sure there isn't anything bent. It is usually a lot easier and cheaper to replace a bent part rather than replace the whole thing after the bent component has destroyed the rest.

Other than the obvious advice (avoid hard landings), there is not much more to be said about the landing gear struts than what is good for the tires and brakes is also good for the struts.

PROPELLERS

Anything said about the propeller also has a great deal to do with the safety of the airplane. Getting that nick in the blade dressed down by your A&P is not only a lot cheaper than buying a new one, but is a lot more pleasant than trying to fly the airplane with two-thirds of a propeller.

Watch out where you do your run-ups. Years ago, when airports were mostly dirt strips, everyone took care, but in these days of paved airports, pilots aren't as diligent as they once were. Avoid gravelly or sandy areas, where you might damage or sand down the propeller. At the same time, make sure you face into the wind; this will tend to minimize the dirt kicked up, as well as keep the engine cool.

This is a good place to remind readers that constant-speed propellers have a fixed life and TBO, just like the engine, but it is unlikely to be the same as your engine's.

THE ELECTRICAL SYSTEM

The electrical system has benefitted greatly by modern electronics technology. The alternator was made possible by this technology, as was the solid-state regulator, which has eliminated most of the problems with the old mechanical type. (Alas, it has also introduced some new ones all of its own.) In addition, the loads imposed upon the electrical system by the avionics have been greatly reduced by the transistorization of the avionics. In fact, the modern small airplane couldn't carry either in volume or weight the avionics it does today if they were the 1950s equivalents. In addition, the whole panel is likely to consume less current than *one* of the old communications sets!

Although the FAA does not let us do our own work on the electrical system, we *are* allowed to work on the battery, and the airplane is fitted with an ammeter, which allows us to do almost all of our own troubleshooting. All electrical system trouble can generally be diagnosed by using these two tools. The system outlined here is much like the troubleshooting outlined in the avionics section, except much simpler.

The Ammeter

One of the reasons that the electrical system is so much easier to troubleshoot than the avionics is that the main troubleshooting instrument is located right in the instrument panel, in the form of the humble ammeter. I am consistently surprised by the number of pilots who use the ammeter as an "on-off" gauge, and who are not aware of the subtleties of its use.

Ammeters come in two flavors—the most usual sort with a center zero, showing charging on the right side (or green zone), and discharge on the left side (or red zone). Another sort frequently used these days has the zero on the left side, and shows only the percentage of the total alternator/generator capacity being used. (From here on, the alternator or generator will be referred to as the generator, for purposes of simplicity.)

The center zero ammeter indicates, on the positive, or charge side, how much current is being sent *to* the battery by the generator. The discharge side indicates how much current is being drawn *from* the battery. The amounts are governed by the regulator, which senses the charge state of the battery and regulates the production of current in the electrical system, as well as the voltage.

When an airplane engine is started, the starter draws a considerable amount of current, which leaves the battery in a discharged state. This current must be replaced by the generator as the engine is operated. Therefore, immediately after starting the engine and for some time thereafter, the generator will be generating a large amount of current, over and above the system's operating needs, to charge the battery. This will be indicated by a large deflection of the needle into the positive side of the ammeter (Fig. 8-1). When the battery has recovered from the load of starting, the generator will

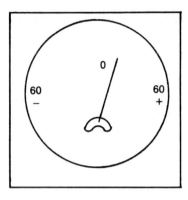

Fig. 8-1. Ammeter making up for starting load, or other major load.

cease generating this current, and the needle will return to zero, showing the battery charged, and the generator generating sufficient current to operate the electrical accessories (Fig. 8-2). This happens slowly, as the regulator "tapers" the charge as the battery approaches full charge. The needle's return to zero shows that the system is in balance. (See Fig. 8-3).

But the needle isn't in the center, you say. Well, the reason for this is that this deflection shows the slight "trickle" charge that the battery is getting while the generator is operating. (On airplanes that have one magneto and one battery ignition system, the deflection will be greater, as it will also show the amount of current generated to operate the ignition system.)

With a little bit of operation of your airplane, you should have a pretty good idea of just how long it takes the electrical system to recover from the average start, as well as from a hard start. (It will take longer on a hard start, as more current has been consumed from the battery, and therefore must be replaced.) It is important to have a good idea of this recovery time, as a constant positive charge over

Fig. 8-2. Ammeter at "neutral position" (slightly in "plus" side).

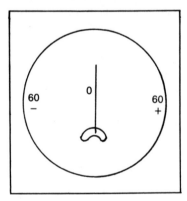

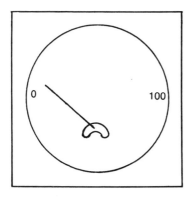

Fig. 8-3. Loadmeter, which reads in percentages of total available.

the small normal deflection shows that the generator is overcharging the battery, or the battery is defective and is not taking the charge. This is usually the fault of the regulator, which instructs the generator what to do in this area. If the battery is being overcharged, it will usually show signs of this by spattering acid out of the vent caps, as well as running much hotter than normal. Overcharging will kill a battery very quickly, so it is important to have the regulator fixed *before* it does this, as soon as the condition is noticed. If the battery is defective (in addition to the obvious danger of operating an airplane with a primary system that is faulty), it is hard on the generator and causes excessive fuel consumption.

Once the electrical system is running at its normal charge, the best way to check the system for proper capacity and operation is to load it. This is done by turning on the high-current-consuming landing lights and taxiing lights. The needle should jump momentarily when the lights are switched on, returning to the "neutral" position immediately. If it fails to do this, if it shows a heavy or partial discharge, this indicates a problem in the generator or regulator. The airplane should be grounded until the problem is repaired, as an airplane battery will not keep the electrical system operating very long.

Since the demands of the electrical system are not allowed (by the FAA) to exceed the generator capacity of the system, this situation will only occur when the generator or regulator is faulty or there is an excessive load being applied to the system by a defective component. A defective component that is consuming excessive current is turning it into heat, and heat causes fire. In addition to the danger of fire, a problem of this nature can cause damage to the electrical system, alternator or regulator. If you suspect an individual component, rather than the generator or regulator, turn off

149

the suspected components to isolate the correct one. If this doesn't locate it, the problem could still be in the wiring or in an electrical component that cannot be easily turned off.

Strobe lights and other heavy intermittent users of current will generally show up in the ammeter readings as little jumps of the needle (as if you had turned on the landing lights), but they should not cause the ammeter reading to go on the discharge side.

In other words, the zero center ammeter is an indicator of battery condition, alternator, and regulator function as well as a monitor of current use and a harbinger of problems to come in the electrical system. Not bad for one little instrument. With proper use, it will allow you to avoid overcharging your battery or over-loading your generator, and will generally allow you to correct electrical problems before they get expensive.

This can save you a considerable amount of money, as usually these components are allowed to continue their errant conduct until the battery or alternator is destroyed and needs replacement. All this, of course, presupposes that you glance at the ammeter from time to time as part of your instrument scan to familiarize yourself with "normal" readings, as well as catch abnormal ones before they cause expensive damage.

There are other varieties of ammeters. The loadmeter, which has zero on the left (Fig. 8-3) is one of these. It shows the total current being consumed by the system, the total load being de-manded of the generator. These, since they deliver very little useful information on their own, are combined with little warning lights that show under and over-voltage conditions to tell you (after the fact) that the generator has failed or the field of the alternator has been shut down to avoid destruction in an over-voltage situation.

Some airplanes equipped with zero center ammeters also have an over-voltage cut-out to remove the field from the alternator in case of an over-voltage condition. If the warning light comes on, and the ammeter shows a discharge situation, turn off the avionics and cycle the master switch to reset the cut-out. This will usually correct the problem. High voltage spikes, too short to damage the general electrical system and components, can damage the alter-nator, which is why the cut-out is installed. An occasional tripping of the over-voltage cut-out shouldn't be worried about, but if it starts tripping consistently, have it checked by your A&P.

Alternator/Generator

Alternators and generators cannot be overloaded as long as their regulators are in working order. Mechanical regulators used on dc generators are prone to allowing overcharging, but this can be spotted easily by scanning the ammeter. Transistorized regulators can also cause this condition, but infrequently. As rare as this problem is, it causes a great deal of very expensive and potentially dangerous trouble if not nipped in the bud.

The battery is overcharged and frequently damaged by a generator run amok, and under certain conditions the generator can be overloaded, ruining it. In addition, depending upon the severity of the problem, the resulting high voltage in the system can destroy radios, lights, and other components of the system. There is no pilot maintenance of the electrical system, but careful monitoring of the ammeter will allow the pilot to catch most electrical problems before they get serious and expensive.

Starter

The starter is just like the one in your car, unless you have an older airplane, in which case it may be of the extra low maintenance "Armstrong" variety, or any number of strange, high maintenance devices thought up to turn over large engines with minimum energy expenditure and weight. In any case, the best advice to prolong its longevity is to use it as little as possible. In other words, keep the engine in top shape so it starts readily, warm the engine and use external power to start the engine when it is cold, and if it doesn't start readily, severely limit the continual use of the starter. It is not very often I recommend not paying attention to your airplane's manual, but most of them recommend maximum continuous use of the starter for 50 seconds. Cut this down to 5-10 seconds, and your generator will repay you with a considerably extended life. All the things that are good for the engine and battery are also good for the starter.

Don't operate the starter when you know the brushes are bad or if there is some other problem with it. It is cheaper to replace brushes than the whole starter, and bad brushes will soon ruin the armature if they are not replaced. A bad Bendix drive or solenoid can ruin the engaging gear as well as hang up the starter instead of allowing it to disengage once the engine is started. This does a starter in more quickly than anything else, and obviously should be avoided. (Some airplanes now have a warning light to warn of this condition.)

THE INSTRUMENTS AND VACUUM SYSTEM

Instruments don't require much preventive maintenance. As a

matter of fact, short of breaking the glass or mechanical knobs on the front, there is not much that even the most ham-handed operator can do to them. If a gyro instrument starts making terminal noises, taking care of it before it fails can both save money and add immeasurably to your safety.

When cleaning the lenses, spray cleaning solution on the rag, not on the instrument face. This avoids getting moisture inside. And of course you should make an effort to keep water out of the backside, as recommended earlier.

The vacuum system is vitally important, but not frequently thought about by the average pilot. (If you do a lot of nighttime IFR or mountain flying, you think about it all the time!) There is an intake filter, which should be attended to at the recommended intervals, *more often* if you are a smoker or allow people to smoke in the airplane.

A little safety note here: vacuum pumps have a limited life— generally about 500 hours at present. If you do a lot of hard IFR, 300 hours might be a more prudent interval. In making your decision, consider how well you do on needle, ball, and airspeed, particularly if you have been following a tumbling gyro.

TURBOCHARGERS

Turbochargers are remarkably hardy. They don't of their own volition wear out; if properly operated, they seemingly will last forever. Proper operation means allowing the oil pressure in the engine to come up to normal before advancing the throttle past low idle position, as well as allowing the turbocharger to cool for a few minutes before shutting it down. Both these techniques will also add hours to your engine, but for the turbocharger they are vital, as the turbocharger's weak spot is the lubrication of the bearing on the compressor-turbine shaft, and these techniques obviate any problems from this bearing.

HANGARS

The temperatures of an airplane parked in the sun on an 85-degree day can reach 160 degrees. Once the heat has baked your airplane thoroughly, it is time for the rain to come and flood the instruments and interior that you haven't quite gotten around to waterproofing as recommended earlier. The next day, we can then poach the interior; and how about some sleet, snow, hail, and a windstorm or two to round things out? Twelve hours out of every 24, ultraviolet rays will help in the deterioration of your once-proud bird by doing their best to make a mess out of the plexiglas plastic, vinyl, rubber, and paint.

A superhuman application of the methods described previously will dull the worst ravages of this mistreatment, but the obvious answer in a hanger. On the other hand, they are expensive, and there probably isn't any hangar space available at your local aerodrome anyhow. Before deciding that hangars are expensive, it might be a good idea to get a handle on what the costs really are. The question of their availability will be left for later.

In figuring hangaring costs, remember your airplane probably cost about as much as a Rolls Royce, and you certainly wouldn't treat one of *them* that way. The benefits of hangaring may be obvious, but computing their exact financial cost is very difficult. There are now available in most parts of the country portable, prefabricated hangars suitable for one airplane, for about ten percent of the cost of a medium-performance single-engine airplane. When figuring direct monetary cost, you should not forget the ten percent reduction in premiums that most insurers allow for hangaring. So much for the easy figures.

It is impossible to compute the savings in maintenance, not only for the airframe, but the engine, avionics, and accessories. An airplane will look better longer when hangared, no matter how carefully you maintain it. Additionally, losses from theft have been shown to be reduced, as well as damage from salt-laden moisture in seashore areas, and industrial pollution in urban areas.

None of this, of course, affects you if you cannot get hangar space at your local airport. If you decide you really want a hangar, it might not be as hard as you think. You may be able to convince the management to add hangar capacity, either prefabricated or not, or if that is impossible, you might be able to get a long-term lease on a space upon which you could erect your own portable hangar. The same sort of imagination that helped you get your license, buy your airplane, and continue paying the bills can also get you a hangar, if you really want one.

Index

Edited by Steven Mesner